ANALYSIS AND DESIGN OF ALGORITHMS FOR COMBINATORIAL PROBLEMS

Edited by

G. AUSIELLO
University of Rome
Italy

and

M. LUCERTINI
University of Rome and IASI-CNR
Rome, Italy

Sponsored by the Institute for System Analysis and Informatics
of the Italian National Research Council (IASI-CNR)

1985

NORTH-HOLLAND – AMSTERDAM · NEW YORK · OXFORD

ISBN: 0 444 87699 5

Publishers:

ELSEVIER SCIENCE PUBLISHERS B.V.
P.O. Box 1991
1000 BZ Amsterdam
The Netherlands

Sole distributors for the U.S.A. and Canada:

ELSEVIER SCIENCE PUBLISHING COMPANY, INC.
52 Vanderbilt Avenue
New York, N.Y. 10017
U.S.A.

Library of Congress Cataloging in Publication Data

Main entry under title:

Analysis and design of algorithms for combinatorial
   problems.

   (Annals of discrete mathematics ; 25) (North-Holland
mathematics studies ; 109)
   "Sponsored by the Institute for System Analysis and
Informatics of the Italian National Research Council
(IASI-CNR)"
   A selected collection of papers based on the workshop
held at the International Centre for Mechanical Sciences
(CISM) in Udine, Italy, in Sept. 1982.
   1. Combinatorial analysis--Data processing--Addresses,
essays, lectures. I. Ausiello, G. (Giorgio), 1941-
II. Lucertini, M. (Mario) III. Istituto di analisi dei
sistemi ed informatica (Italy) IV. Series. V. Series:
North-Holland mathematics studies ; 109.
QA164.A49 1985   511'.6   84-28667
ISBN 0-444-87699-5

PRINTED IN THE NETHERLANDS

FOREWORD

This volume contains a selected collection of papers based on the issues presented by the authors at the Workshop "Analysis and design of algorithms for combinatorial problems", held at CISM in Udine in September 1982.

Combinatorial problems are from the very beginning in the history of mathematics; maze threading and the Konisberg bridges crossing are only two examples of early combinatorial problems. Up to the sixties the main classes of combinatorial problems, linear and integer programming problems and some basic solution techniques (like branch and bound, dynamic programming, cutting planes, etc.), have been defined, often in connection with some applications. In particular, during the sixties, a great amount of research contributions has been produced in graph theory, laying the foundations for most of the research in graph optimization of the following years.

During the seventies a large number of special purpose models have been developed and the related applications deeply analyzed (such as various kind of scheduling models, assignment models, network flow models, etc.). In the same time many algorithmic approaches have been introduced based on different philosophies and evaluated from a computational viewpoint in completely different ways. During the same period many researchers started working on the theoretical basis of combinatorial algorithms theory in order to integrate the different approaches in few basic concepts and to point out new criteria for the design and the computational analysis of combinatorial algorithms. The NP–completeness theory and the related classifications of combinatorial problems have been mainly developed in this period; the definitions of "ε-approximate algorithms", "fully polynomial-time approximation scheme", the theory behind the local search methods, the basic algorithmic results on polyhedral combinatorics, perfect graphs and matroids, the Lagrangean duality theory in integer programming, have all been developed in the seventies (and most of them are still hot research issues).

The impressive growth of the field has been strongly determined by the demand of applications and influenced by the technological increases in computing power and the availability of data and software. In fact the availability of such basic tools have led to the feasbility of the exact or well approximate solution of large scale realistic combinatorial optimization problems (and have created a number of new combinatorial problems).

Such developments of research on algorithmic aspects of combinatorial problems has led to a fruitful cooperation between experts in discrete mathematics, in theoretical computer science and in operations research, with the aim of building a common background for researchers in the field. A pioneering step in this direction may be ascribed to the Conference on "Interfaces between Computer Science and Operations Research" which was held at the Mathematische Centrum in Amsterdam in 1978. One year later the School on "Analysis and Design of Algorithms in Combinatorial Optimization" was held at CISM in Udine (1) and was followed by the School on "Combinatorics and Complexity of Algorithms" (Barcelona, 1980). The workshop which provided the basis for this volume was held at CISM in Udine, in 1982 and was subsequently followed by a School on "Algorithms Design for Computer System Design" again at CISM, in Udine (2), in July 1983.

The subjects of the papers collected in this volume range from optimization over graphs and hypergraphs to basic combinatorics, from data structures to machine models and from probabilistic analysis of algorithms to applications in computer architecture design.

We wish to thank the Department of Computer and System Science of the University of Rome and the CISM in Udine for the contribution given in organizing the Workshop. This volume is due to the

---

(1) G. Ausiello, M. Lucertini (Eds.): "Analysis and design of algorithms in combinatorial optimization", CISM Courses and Lectures N. 266, Springer-Verlag, New York, 1981.

(2) G. Ausiello, M. Lucertini, P. Serafini (Eds.): "Algorithm design for computer system design" CISM Courses and Lectures N.284, Springer-Verlag, New York, 1984.

generous support of the Institute of System Analysis and Informatics of the Italian National Research Council; to the director of the Institute prof. L. Bianco, we express our gratitude. Sincere thanks are also due to all referees who have given their contribution to the selection and improvement of the papers contained in the volume.

Rome, September 1984.

Giorgio AUSIELLO

Mario LUCERTINI

CONTENTS

Annals of Discrete Mathematics 25 (1985) 1-26

STRONGLY EQUIVALENT DIRECTED HYPERGRAPHS

G. AUSIELLO, A. D'ATRI

University "La Sapienza" Rome, Italy

D. SACCA'

CRAI Rende, Italy

Various concepts of equivalence among graphs and among hypergraphs have been studied by several authors. In this paper the concept of strong equivalence between directed hypergraphs, based on the notion of kernel, is considered and its properties are analyzed. The computational complexity of testing strong equivalence and of determining a minimal strongly equivalent hypergraph is examined and algorithms which solve such problems in polynomial time are provided.

1. Introduction

Directed hypergraphs are a generalization of the concept of directed graphs which have been studied, in various forms, for modelling problems in computer science [2, 3, 5, 7, 9, 11].

Authors' addresses:

Giorgio Ausiello and Alessandro D'Atri: Dipartimento di Informatica e Sistemistica, Via Buonarroti 12, I–00185, Roma, Italy.

Domenico Sacca': Consorzio per la Ricerca e le Applicazioni di Informatica, Via Bernini 5, I–87036, Quattromiglia di Rende, (CS), Italy.

Categories and Subject Descriptors:
G.2.2 [Discrete Mathematics] : Graph Theory-*graph algorithms:*

General Terms: Algorithms, Theory

Additional Key Words and Phrases: closure, hypergraph, kernel, NP–complete, polinomial algorithms.

In particular in [2] the notion of equivalence between directed hypergraphs has been introduced together with some of its applications. Such a notion is based on the generalization to directed hypergraphs of the concepts of traversal and of transitive closure of graphs, and states that (analogously to the case of graphs) two hypergraphs are equivalent if they have the same closure.

In this paper a stronger notion of equivalence is introduced, based on the concept of kernel of a hypergraph. The kernel K of a hypergraph **H** is a minimal family of source sets of **H** (a source set is a set of nodes which is the source of at least one hyperarc) such that each source set in **H** may be reached from a source set in K. Two equivalent hypergraphs are said to be strongly equivalent if their kernels are mutually related with respect to their closures.

A meaningful application of the strong equivalence among hypergraphs may be found in relational database theory [8,10]. In such a context, hypergraphs may be used to model a set of integrity constraints and the strong equivalence of two hypergraphs represents the property that the corresponding constraints allow the same set of legal instances of the data-base.

The aim of this paper is to study the structural and the computational properties of strongly equivalent hypergraphs. In the next paragraph the fundamental definitions of hypergraph, hypergraph closure and kernel of a hypergraph are introduced and their basic properties are shown. In particular it is shown that both the closure of the source sets and a minimum kernel of a given hypergraph (that is a kernel composed by the minimum number of source sets) may be found in polynomial time. In Paragraph 3 the problem of deciding the strong equivalence of two hypergraphs is approached and an algorithm which solves such a problem in time quadratic in the length of the hypergraph representation is provided. Finally in Paragraph 4 strongly equivalent minimal representations of hypergraphs are studied and it is shown that the problem of determining a strongly equivalent hypergraph with the minimum number of source sets is also solvable in polynomial time, while all other minimization problems are NP–complete.

2. Directed hypergraphs and their kernels

Hypergraphs [4] are widely used for representing structures and concepts in several areas of computer science. In this paper we shall deal with a particular class of hypergraphs where a hyperarc is an ordered pair composed by a set of nodes and a single node.

A *directed hypergraph* **H** is a pair $<N, H>$, where N is the set of *nodes* and H is a set of pairs (X, i), called *hyperarcs,* such that i is a node in N and X a subset of N. We define a *source set* to be a set of nodes that appear as left side of at least one hyperarc.

From now on we shall refer to directed hypergraphs simply as hypergraphs.

Example 1. The hypergraph **H** $= <\{A, B, C, D, E, F\}, \{(\{A, B\}, C),$ $(\{B\}, D), (\{C, D\}, E), (\{C, D\}, F)\}>$ is shown in Fig. 1, where hyperarcs are represented by arrows. The source sets of **H** are $\{A, B\}, \{B\}$ and $\{C, D\}$.

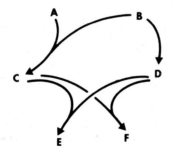

Figure 1. A directed hypergraph

We shall denote by n the number of nodes of the hypergraph, by m the number of hyperarcs, by s the number of source sets, and by a the sum of the cardinalities of all source sets *(source area)*. Finally, we refer to the overall length of description of the hypergraph by $|\mathbf{H}|$.

In the previous example we have:

$$n = 6, \quad m = 4, \quad s = 3, \quad a = 5$$

As far as the length of the description is concerned, if we assume a representation based on adjacency lists (where for every source set the list of adjacent nodes is given) we have $|H| \cong a + m$. According to the same representation the number of source sets, s, corresponds to the number of adjacency lists.

In this paper we are mainly concerned in discussing the equivalence between hypergraphs, based on the concepts of traversal and closure of hypergraphs (as it will be defined formally by extending the corresponding concepts for ordinary directed graphs).

Given a hypergraph $H = <N, H>$, the *closure* of H, denoted by H^+, is the hypergraph $<N, H^+>$ such that (X, i) is in H^+ if one of the following conditions holds:

— (X, i) is a hyperarc in H, or

— i is an element of X *(extended reflexivity)*, or

— there exists a set of nodes $Y = \{n_1, \ldots, n_q\}$ such that there exist hyperarcs (X, n_j), for $j = 1, \ldots, q$, in H^+ and (Y, i) is a hyperarc in H *(extended transitivity)*.

Note that when both X and Y are singletons, the extended reflexivity and the extended transitivity coincide with the usual notions of reflexivity and transitivity as they are defined on graphs. Given the hypergraph H in Fig. 1 some of the hyperarcs in H^+ are: ({A, B}, C), ({A, B}, A), ({A, B}, E).

Given a source set X of a hypergraph $H = <N, H>$, we call *closure of X*, denoted X^+, the set of all nodes i in N such that there exists a hyperarc (X, i) in H^+.

Fact 1. Let $H = <N, H>$ be a hypergraph and let S_i be a source set in H. The closure S_i^+ of S_i can be found in linear time.

Proof. Let $\underline{S} = \{S_1, \ldots, S_s\}$ be the set of all source sets in H and let S_i be a source set in $\underline{S}$. For each S_j in $\underline{S}$, we represent all hyperarcs leaving S_j by an adjacency list L_j, which also contains all nodes in S_j. Moreover, for each S_j in $\underline{S}$, different from S_i, we introduce the counter q_j which keeps track of the number of nodes of S_j which are currently in the closure of S_i. Finally, for each node n_r in N, we introduce the list M_r of all source sets, but S_i, containing n_r (actually, we suppose that M_r and all other data structures which are used in the following refer to source sets through their indices instead of their explicit representations). We start by setting S_i^+ to the empty set. Then we insert S_i in an empty list I. For each source set S_j in I we perform the following steps. We first remove S_j from I, and then for each node n_r in L_j which is not already in S_i^+, we add n_r to S_i^+ and for each source set S_k in M_r, we increase the counter q_k by one. As soon as the source counter becomes equal to the cardinality of S_k, we insert S_k in the list I. The previous steps are performed until I becomes empty. It is easy to see that this algorithm correctly computes S_i^+ and that it works in $0(|\mathbf{H}|)$ time.

Q.E.D.

We say that two hypergraphs are *equivalent* if they have the same closure.

In [2] the problem of determining equivalent hypergraphs has been approached. In particular various definitions of minimal representations of equivalent hypergraphs have been given and the complexity of the corresponding minimization problems has been analyzed.

In this paper we deal with a stronger definition of hypergraph equivalence based on the structural properties of the kernels of two hypergraphs.

We remind that, given a directed graph $G = <N, A>$, a *kernel* of G is defined to be a subset K of N such that no two nodes in K are joined by an arc in A and such that for each node i in $N \setminus K$ there is a node j in K for which (j, i) belongs to A.

Note that not all graphs have a kernel and that the problem of determining whether a graph has a kernel or not is NP–complete [6]. Nevertheless, if we consider a graph G closed under transitivity it is easy to see that a kernel of G always exists and may be found in time linear in the number of arcs. Starting from this consideration we define a kernel of a hypergraph by referring to the hyperarcs in the closure.

A *kernel of a hypergraph* $H = \langle N, H \rangle$ is a minimal family of source sets $K = \{S_1, \ldots, S_k\}$ such that for each source set X in H there is a source set S_i in K for which X is contained in S_i^+.

Example 2. The hypergraph H' in Fig. 2a) has only one kernel that is $K = \{\{A, B\}\}, \{B, C\}, \{E, D\}\}$. The hypergraph H in Fig. 2b) has four kernels, $K_1 = \{\{A, B\}\}$, $K_2 = \{\{A, B, C\}\}$, $K_3 = \{\{C\}\}$, and $K_4 = \{\{B\}\}$.

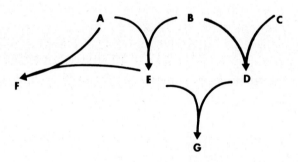

a) A hypergraph H which has only one kernel

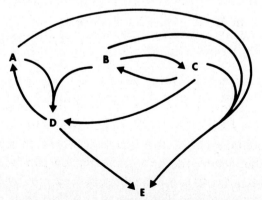

b) A hypergraph H' which has four kernels

Figure 2.

Let us now show how to find a kernel of a hypergraph.

Proposition 1. A kernel of a hypergraph H can be found in $O(s \cdot |H|)$ time where s is the number of source sets in H and $|H|$ the length of the description of H.

Proof. Let $\underline{S} = \{S_1, \ldots, S_s\}$ be the family of all source sets in H. For each source set S_i in S, we compute the closure S_i^+ by the algorithm given in the proof of Fact 1. Moreover for each source set S_j such that S_j is contained in S_i^+, we remove S_j from $\underline{S}$. We repeat this step until we compute the closure of all source sets in $\underline{S}$. At the end $\underline{S}$ will contain all source sets of a kernel of H. Clearly the algorithm runs in $O(s \cdot |H|)$ time, since this is the time needed to perform the closure of all source sets in H.

Q.E.D.

Given two hypergraphs $H = <N, H>$ and $H' = <N, H'>$, we say that a kernel K of H is *equivalent* to a kernel K' of H' if for each source set S_i in K there exists a source set S_j' in K' such that $S_i^+ = (S_j')^+$ and conversely.

Theorem 1. All kernels of a hypergraph are pairwise equivalent and consist of the same number of source sets.

Proof. Let $H = <N, H>$ be a hypergraph an let R be the set of all source sets S_i in H such that there exists no source set S_j in H for which S_i^+ is properly contained in S_j^+. We define an equivalence relation in R as follows. Two source sets S_i and S_j in R are equivalent if $S_i^+ = S_j^+$. Let us now consider any kernel K of H. By definition of kernel and by costruction of R, K contains at least one source set in each equivalence class of R. Moreover, since we require that a kernel be minimal, K contains exactly one source set in each equivalence class of R and no other source set is in K. Let us now assume that H has a distinct kernel K'; by repeating the previous argument, K' has to contain exactly one source set in each equivalence class in R. Hence for each source set S_i in K, there is one and only one source set S_j in K' such that $S_i^+ = S_j^+$. Therefore, it follows that K and K' are equivalent and have the same number of source sets.

Q.E.D.

Theorem 1 says that all kernels of a hypergraph are composed by the same number of source sets. Actually, we are also interested in finding a kernel with the minimum source area, that is a kernel for which the sum of the cardinalities of its source sets is minimum. We call such a kernel an *optimum kernel*. For instance the hypergraph **H'** in Fig. 2b) has two optimum kernels $K_3 = \{\{C\}\}$, and $K_4 = \{\{B\}\}$.

Proposition 2. An optimum kernel of a hypergraph **H** can be found in $O(s \cdot |H|)$ time, where s is the number of source sets in **H** and $|H|$ the length of the description of **H**.

Proof. We start from an empty set K and we compute the closure of all source sets in K in $O(s \cdot |H|)$ time. Then we construct the directed graph $G = \langle V, E \rangle$, where each node i in V corresponds to the source set S_i in **H** and there is an arc (i, j) in E if and only if S_j is contained in S_i^+, where S_i and S_j are the source sets of **H** corresponding to i and j respectively. Since G is obviously closed under transitivity we can compute the strong components of G in $O(s^2)$ time (where s is equal to the number of nodes in the graph G). Now, let C_k be a strong component of G such that there exists no arc (i, j) in E where j is in C_k and i is not in C_k. For each such component C_k, we insert in k a source set with minimum cardinality among all source sets corresponding to the nodes in C_k. This step can be performed in $O(a)$ time, where a is the source area of **H**. It is easy to see that the set K is a optimum kernel of **H**. The whole algorithm runs in $O(s \cdot |H|)$ time since $s \le |H|$ and $a \le |H|$.

Q.E.D.

3. Strongly equivalent hypergraphs

In this section we introduce the notion of strong equivalence among hypergraphs.

We say that two hypergraphs are *strongly equivalent* if they are equivalent and their kernels are equivalent.

Notice that, in general, the equivalence of two hypergraphs does not imply their strong equivalence. However, in the case the two hypergraphs coincide with ordinary graphs if they are equivalent they are also strongly equivalent.

Example 3. Let us consider the three hypergraphs H_1, H_2 and H_3 given in Fig. 3.

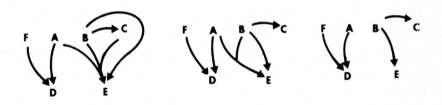

a) A hypergraph H_1 b) A hypergraph H_2 c) A hypergraph H_3

Figure 3. Equivalent hypergraphs

While both H_2 and H_3 are equivalent to H_1, only H_2 is strongly equivalent to H_1. Let us now consider the hypergraph H' given in Fig. 2b), the hypergraph H in Fig. 4 is strongly equivalent to H'.

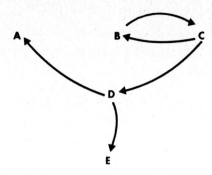

Figure 4. A hypergraph H strongly equivalent to the hypergraph of Fig. 2b).

The introduction of such a stronger notion of equivalence between hypergraphs derives not only from the interest of studying structural properties of directed hypergraphs, but also from the fact that the equivalence of kernels has a meaningful interpretation in various application. In particular, in relational database theory directed hypergraphs may be used for representing, under suitable assumptions, database schemes described by a full join dependency and a set of functional dependencies. In this context, it is possible to show that two database schemes are equivalent (that is they allow the same set of legal instances) if the corresponding hypergraphs are strongly equivalent.

A property of strongly equivalent hypergraphs which directly derives from the definitions of strong equivalence and equivalent kernels is the following.

Proposition 3. The kernels of two strongly equivalent hypergraphs have the same cardinality.

Proof. Let **H** and **H'** be two strongly equivalent hypergraphs. Let K be a kernel of **H** and K' be a kernel of **H'**. Since all kernels of **H** (**H'**) have the same cardinality of K (K') by Theorem 1, we only have to prove that K and K' have the same cardinality. By definition of strong equivalence, K and K' are equivalent, i.e., there is a mapping f : K $\longrightarrow$ K' such that for each S_i in K, we have $S_i^+ = (f(S_i))^+$. The mapping f is injective because otherwise two source sets in K would have the same closure (contradiction with the fact that, by definition, no proper subset of K is a kernel). Moreover, the mapping f is also surjective since, by definition of equivalent kernels, for each S_i' in K', there is S_i in K for which $(S_i')^+ = S_i^+$. It follows that f is bijective and, then, K and K' have the same cardinality.

Q.E.D.

In order to present an algorithm for deciding whether two hypergraphs are strongly equivalent or not, we need to prove the following important property of strongly equivalent hypergraphs.

Theorem 2. Two hypergraphs **H** $= <$N, H$>$ and **H'** $= <$N, H'$>$ are strongly equivalent if and only if for each hyperarc (X, i) in **H** such a hyperarc is also in $\mathbf{H'}^+$ and there exists a source set X' in **H'** such that X is contained in X'^+ (and conversely).

Proof. *If part.* Let us suppose that for each hyperarc (X, i) in **H**, there exists a source set X' in **H'** such that X is contained in X^+ and conversely. In this case, by definition of kernel, **H** and **H'** have equivalent kernels. Let us now also suppose that for each hyperarc (X, i) in H, the hyperarc (X, i) is also in H^+ and conversely. By definition of strong equivalence, in order to conclude the proof, we have to show that **H** and **H'** are equivalent. We have that H is contained in H^+ by assumption. Hence, considering the definition of hypergraph closure, H^+ is contained in H^{*+}; by exchanging H with H' we also have H^+ contained in H^*. It follows that **H** and **H'** are strongly equivalent.

Only if part. Let us suppose that **H** and **H'** are strongly equivalent, that is that they are equivalent and have equivalent kernels. Let us consider any hyperarc (X, i) in **H**. This hyperarc is also in H^+, since $H^+ = H'^+$ by hypothesis. Let us now prove that there exists a source set X' in **H'** such that X is contained in X^+. By definition of kernel there exists a source set S_i in a kernel of **H** (say K), such that X is contained in S_i^+. Let us consider a kernel K' of H'. Since K and K' are equivalent, there exists a source set S_i' in K' such that $S_i^+ = (S_i')^+$. Hence X is contained in $S_i'^+$. By exchanging H with H' and viceversa in the above argument, we also have that for each (X', i) in H', (X', i) is also in H^+ and there exists a source set X in H such that X' is contained in X^+. This concludes the proof.

Q.E.D.

Theorem 2 suggests a simple polynomial time algorithm for testing the strong equivalence between two hypergraphs.

Proposition 4. Let $H = <N, H>$ and $H' = <N, H'>$ be two hypergraphs. The recognition of whether H and H' are strongly equivalent or not can be done in $0((s + s') \cdot (|H| + |H'|))$ time, where s and s' are the number of source sets respectively in H and in H', and |H| and |H'| are the lengths of the description of H and H' respectively.

Proof. Let us consider the following algorithm. Let R' be the family of all source sets in H' which are not in H. For each source set X' in R' we choose a node i in it. We consider the hypergraph $H_1 = <N, H_1>$, where H_1 is obtained by adding to the hyperarcs in H the hyperarcs (X', i) such that X' is in R' and i is the node in X' we have previously choosen. We compute the closure of all source sets in H_1 in $0((s + s') \cdot (|H| + a'))$ time, where a' is the size of the source area of H' (see Fact 1). Similarly, we costruct a hypergraph H_1' starting from H' and the family of source sets of H not in H'. We compute the closure of all source sets of H_1 in $0((s + s') \cdot (|H'| + a))$ time, where a is the size of the source area of H. By Theorem 2, the hypergraphs H and H' are strongly equivalent if and only if

a)   for each (X', i) in H', (X', i) is in H_1^+, (thus i is in the closure of X' in H_1);

b)   for each X' in S', there exists a source set X in H_1, but not in S', such that X' is contained in X^+ (and conversely).

We note that the two conditions can be easily checked while the closures of the source sets in H_1 and H_1' are computed and without increasing the complexity of these computations. Hence the recognition of whether H and H' are strongly equivalent or not can be done in $O((s + s') \cdot (|H| + |H'| + a + a'))$ time and, hence, in $O((s + s') \cdot (|H| + |H'|))$ time, since obviously $a \leq |H|$ and $a' \leq |H'|$.

Q.E.D.

From the definition of equivalent kernels and strongly equivalent hypergraphs it follows that the kernels of two strongly equivalent hypergraphs have the same cardinality; however they may have different source area. A problem is to find a strongly equivalent hypergraph whose optimum kernel has the minimum source area (i.e., no optimum kernel of any strongly equivalent hypergraph has a smaller source area). We shall prove that this problem cannot be solved in polynomial time (unless $P = NP$).

Example 4.   The hypergraph H in Fig. 5a) has only one kernel $K = \{\{A, B, C\}\}$. The hypergraph H' in Fig. 5b) is strongly equivalent to H and has the kernel $K = \{\{A, D\}\}$ with the minimum source area.

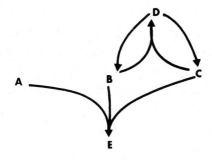

a) A hypergraph H

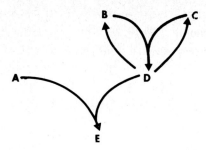

b) A hypergraph H' strongly equivalent to H having a kernel with
the minimum source area.

Figure 5

Theorem 3. Given a hypergraph **H** and an integer b, the problem of finding a hypergraph **H'** strongly equivalent to **H** with a kernel wich has the source area not greater than b is NP—complete.

Proof. Let us first prove that the problem is in NP and then that it is NP—complete. Let $H = <N, H>$ be a hypergraph and let $K = \{ S_1, ... S_m \}$ be a kernel of **H**. In order to prove that our problem is in NP, we costruct **H'** as follows. For each S_i in K we guess a set of nodes $\underline{S}_i$ such that $S_i^+ = \underline{S}_i$ in **H** and the sum of the cardinalities of all $\underline{S}_i$ is not greater then b. Moreover, we add the hyperarc $(\underline{S}_i, j)$ to **H**, where j is one of the nodes in $\underline{S}_i$. The hypergraph **H'** obtained in this way is obviously strongly equivalent to **H** since we have added hyperarcs which are in the closure of **H** by reflexivity rule. By construction of **H'**, a kernel of **H'** is $K' = \{ \underline{S}_1, ..., \underline{S}_m \}$. Hence we have shown that a hypergraph **H'** strongly equivalent to **H** with a kernel bounded by b can be found in nondeterministic polynomial time (if it exists).

Let us now prove that the problem is NP—complete. To this end we give a polynomial reduction from the SET_COVERING problem [6] to the problem of finding a hypergraph **H'** strongly equivalent to the given hypergraph **H** such that one of its kernels has the source area bounded by a given b. Let the following instance of the SET_COVERING problem be given: $E = \{e_1, \ldots, e_n\}$ is the set of elements, $F = \{F_1, \ldots, F_m\}$ is a family of subsets of E such that their union is equal to E and b is a positive integer. The SET_COVERING problem consists in finding a subfamily F' of F which has cardinality not greater than b and such that the union of its elements is still equal to E. Given the above instance, we construct a hypergraph **H** whose nodes are $\underline{e}_1, \ldots, \underline{e}_n, \underline{F}_1, \ldots, \underline{F}_m, \underline{T}_1$ and $\underline{T}_2$. Furthermore, for each e_j in F_i we have a corresponding hyperarc $(\underline{F}_i, \underline{e}_j)$. Finally, there are the hyperarcs $(\{\underline{e}_1, \ldots, \underline{e}_n\}, \underline{F}_i)$, where $1 \le i \le m$, and the hyperarc $(\{\underline{F}_1, \ldots, \underline{F}_m, \underline{T}_2\}, \underline{T}_1)$, (see Fig. 6).

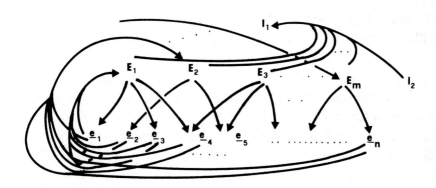

Figure 6. The hypergraph associated to an instance of the set covering problem.

The hypergraph **H** has only one kernel consisting of a single source set $K = \{ \{\underline{F}_1, \ldots, \underline{F}_m, \underline{T}_2\} \}$. Let us now consider a hypergraph **H'** strongly equivalent to **H** and whose kernel K' has source area bounded by b. Since **H** and **H'** have equivalent kernels, by Proposition 3 we have that K' has only one source set, say X'. Moreover, X' may be obtained by adding the node $\underline{T}_2$ to the set $\{\underline{F}_{i1}, \ldots, \underline{F}_{ik}\}$; where for each $\underline{e}_j$ $(1 \le j \le n)$, there is a node $\underline{F}_{ir}$ $(1 \le r \le k)$ such that $(\underline{F}_{ir}, \underline{e}_j)$ is in

H' and the cardinality of X' in minimum, i.e., no other hypergraph **H''** strongly equivalent to **H** has a kernel K'' such that |X''|<|X'|, where X'' is the only source set in K''. Hence, by finding the hypergraph **H'**, we also solve the SET_COVERING problem. It follows that our problem is NP–complete.

Q.E.D.

4. Minimal strongly equivalent hypergraphs

In [2] several problems related with the minimal representation of equivalent hypergraphs (called coverings) have been considered as the extensions of the problem of finding the transitive reduction of a directed graph [1]. In particular minimal coverings of a hypergraph have been defined with respect to various criteria: number of hyperarcs, number of source sets, source area. It has been shown that, in general, the problem of finding such minimal coverings is NP–complete.

The only problem which turns out to be polynomial is the problem of finding a "source minimum" equivalent hypergraph, i.e., a "non-redundant" hypergraph that is equivalent to the given hypergraph and has the minimum number of source sets. In this paper, we show that the corresponding problem in the case of strong equivalence is also solvable in polynomial time.

On the other hand, we shall not treat the corresponding harder problems since, in these cases, the results generalize in a trivial way (actually, the proofs of NP–completeness given in [2] refer to equivalent hypergraphs which are also strongly equivalent).

Let **H** = <N, H> be a hypergraph, we give the following definitions:

— a *subhypergraph* **H'** = <N, H'> of **H** is a hypergraph such that H' is contained in H;

— a hyperarc (X, i) in H is *redundant* if there exists such a hyperarc in the closure of the subhypergraph of **H** obtained by eliminating (X,i) from H:

—   a node j is *redundant* in a source set X of **H** if j is a node in X and there exists in H^+ the hyperarc $(X - \{j\}, j)$;

—   a *strong component* of **H** is a maximal set C_i of source sets in **H** such that all source sets in C_i have the same closure;

—   a source set X is *superfluous* in **H** if there exists another source set Y in the same strong component of X such that if we replace all hyperarcs (X, i) in H by hyperarcs (Y, i), the obtained hypergraph is equivalent to **H** (from now on, we shall refer this hyperarc replacement by "elimination of superfluous source sets").

Finally, let **H** and **H'** be two equivalent hypergraphs and let C_i and C_j' be two strong components of **H** and **H'** respectively. We say that C_i is *equivalent* to C_j' if for any X in C_i there exists X' in C_j' such that $X^+ = X'^+$.

Example 5. Let us consider the hypergraphs H_1 and H_2 of Fig. 3. The hypergraph H_2 is a subhypergraph of H_1. Moreover, the hyperarc $(\{A, B, C\}, E)$ is redundant in H_1 and the node C is redundant in the source set $\{A, B, C\}$ of H_1.

Example 6. Let us now consider the hypergraphs **H** and **H'** of Fig. 7. The strong components of **H** are $C_1 = \{\{A\}, \{B, C\}, \{D, E\}\}$, $C_2 = \{\{G, H, I\}\}$, $C_3 = \{\{B\}, \{D\}\}$, $C_4 = \{\{C\}, \{E\}\}$, $C_5 = \{\{H\}\}$. The source set $\{D, E\}$ in **H** is superfluous and the node I is redundant in the source set $\{G, H, I\}$. The hypergraph **H'** is obtained from **H** by eliminating the superfluous source set $\{D, E\}$ and the redundant node G from the source set $\{G, H, I\}$.

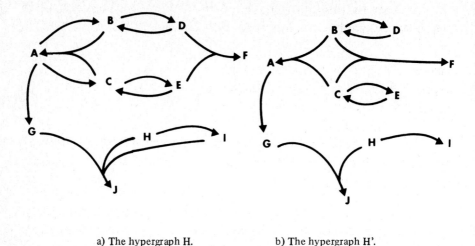

a) The hypergraph H.   b) The hypergraph H'.

Figure 7

Let **H** = <N, H> be a hypergraph. An *SME (Source Minimum Equivalent)* hypergraph of **H** is a hypergraph **H'** = <N, H'> such that:

a)   **H'** is equivalent to **H**,

b)   no source set in **H'** contains redundant nodes,

c)   no proper subhypergraph of **H'** satisfies condition a), and

d)   **H'** has the minimum number of source sets, i.e. there exists no hypergraph which has all previous properties and fewer number of source sets.

If in the above definition the condition a) is replaced by the following condition:

a')   **H'** is strongly equivalent to **H**;

we say that **H'** is an *SMSE (Source Minimum Strongly Equivalent) hypergraph* of **H**.

Note that in this case, the condition c) does not imply that an SMSE—hypergraph has no redundant hyperarcs.

Example 7. Let us consider the hypergraph H_1, H_2 and H_3 of Fig. 3. H_2 is an SME—hypergraph of H_1 but it is not strongly equivalent to H_1. H_3 is an SMSE—hypergraph of H. Note that the hyperarc $\{A, B\}$, E) in H_3 is redundant but it is necessary for H_3 to have the kernel equivalent to the kernel in H_1. Let us now consider the hypergraphs H and H' of Fig. 7. The hypergraph H' is both an SME—hypergraph and a SMSE—hypergraph of H.

Let us now consider the problem of finding an SME—hypergraph of a hypergraph.

Fact 2. An SME—hypergraph of a hypergraph $H = <N, H>$ can be found in $O(|H|^2)$ time.

Proof. Let us consider the following algorithm:

a)   find the closure of all source sets in **H** and all strong components of **H**,

b)   eliminate superfluous source sets,

c)   eliminate redundant hyperarcs,

d)   eliminate redundant nodes in all source sets in **H**.

In [2] it is proved that the algorithm correctly determine an SME— hypergraph in $O(|H|^2)$ time. In particular, we note that step a) can be easily implemented in $O(s \cdot |H|)$ time (see Fact 1). Step b) can be im- plemented as follows (we refer to the data structure given in the proof of Fact 1). For each strong component C_i of **H** and for each S_j in C_i, let us consider any S_k in C_i (which has not been eliminated). We dele- te from the adjacency list L_j all target nodes of the hyperarcs leaving S_j and we add such nodes to L_k. We have that L_j contains all nodes in S_j. We compute the closure of S_j by using the new values of the two adjacency lists L_j and L_k while all other data remain unchanged. S_j is superflous if and only if this closure is equal to the closure of S_j computed in step a). Step b) can be done in $0(s \cdot |H|)$ time, where s is the number of source sets in **H**. Steps c) and d) can be implemented as follows. For each source set S_j in **H**, we consider the adjacency list L_j. For each node in S_j, or hyperarc (S_j, i) in H, we delete i from L_j and compute the closure of S_j. We have that the node i is redundant in S_j or the hyperarc (S_j, i) is redundant in **H** if and only if the new closure of S_j is equal to the closure computed in Step a). These steps can be done in $0(|H|^2)$ time. Hence, the whole algorithm runs in $0(|H|^2)$ time.

Q.E.D.

We give now an important property (proven in [2]) of an SME–hypergraph **H'** of a hypergraph **H** saying that for each strong component of **H'** there exists an equivalent strong component of **H** which has at least the same number of source sets.

Fact 3. Let **H** be a hypergraph and let **H'** be an SME–hypergraph of **H**. Let C and C' be the set of all strong components of respectively **H** and **H'**. There is an injection f: C'--> C such that for each C_i in C', $|C_i'| \leq |C_j|$ and C_i' and C_j are equivalent strong components, where $C_j = f(C_i')$.

We are now ready to present a polynomial time algorithm for finding a SMSE–hypergraph.

Theorem 4. Let $\mathbf{H} = <N, H>$ be a hypergraph. A SMSE–hypergraph of **H** can be found in $0(|H|^2)$ time.

Proof. Let us consider the following algorithm:

a)   Find the closure of all source sets in **H** and determine all strong components of **H**.

b)   Find a kernel K of **H**.

c) Eliminate superfluous source sets from **H**; insert such source sets in a set L.

d) Eliminate redundant hyperarcs from H; if this elimination determines the elimination of source sets insert such source sets in L.

e) For each source set X in L which is also in K, do the following steps:

> Let C_i be the strong component of **H** which contains X. If there exists a source set X'≠X in C_i that is not in L then:

>> e1) replace X by X' in K

> else

>> e2) add the (redundant) hyperarc (X, i), where i is any node in X

f) Eliminate redundant nodes in all source sets of the hypergraph (and in all corresponding source sets in the kernel).

Let **H'** = <N, **H**'> and K' be the hypergraph and the kernel so obtained. We now prove that **H'** is strongly equivalent to **H**.

First of all we note that the set of source sets K' is a kernel of **H'**, by construction. Moreover, **H** and **H'** have equivalent kernels since K and K' are equivalent by construction of K'. Let us now show that **H** and **H'** also have the same closure. By Fact 2, an SME–hypergraph **H"** of **H** can be obtained from **H'** by only deleting the hyperarcs in Step e2). Since these hyperarcs are in **H"**$^+$ by reflexivity, **H'** and **H"** have the same closure. Hence, **H** and **H'** also have the same closure since **H**$^+$=**H"**$^+$ by definition of SME–hypergraph. It follows that **H** and **H'** are strongly equivalent. Let us now prove that **H'** is an SMSE–hypergraph of H. First of all, we note that no hyperarc in **H'** contains redundant nodes because of step f). Moreover, no subhypergraph of **H'** is strongly equivalent to **H'**. In fact, if we eliminate some hyperarcs added by step e2) then the kernel of this subhypergraph is not equivalent to K' and if we eliminate any other hyperarc then the closure of the subhypergraph is not equal to **H'**$^+$ (recall that no subhypergraph of the hypergraph H'' obtained from H' by only deleting the hyperarcs added in step e2) is equivalent to H'' and, then, to H' by definition of SME—hypergraph). Hence, in order to show that H' is an SMSE—hypergraph of H, we only need to prove that H' has the minimum number of source sets. To this end, let us consider an SMSE—hypergraph $\underline{H} = \langle N, \underline{H} \rangle$ of H (this hypergraph obviously exists). We prove that H' and $\underline{H}$ have the same number of source sets. Since $\underline{H}$ is strongly equivalent to H, by assumption, and since we have proved that H' is strongly equivalent to H, H' and $\underline{H}$ are strongly equivalent. This means that K' and $\underline{K}$, where $\underline{K}$ is a kernel of $\underline{H}$, are equivalent and, by Proposition 3, they have the same cardinality. Let us now consider the above defined hypergraph H''. Since H'' is an SME—hypergraph of H, H'' is also an SME—hypergraph of $\underline{H}$. Hence, by Fact 3, there is an injection f' : $C'' \dashrightarrow \underline{C}$ (where C'' ($\underline{C}$) is the set of all strong components of H'' ($\underline{H}$)) such that for each C_i'' in C'' we have $|C_i''| \leq |f'(C_i'')|$ and C_i'' and $f'(C_i'')$ are equivalent. Let us now consider the set C' of all strong components of H'. By costruction of H'', we have that C' is the union of C'' and M', where $M' = \{C_i' | C_i' = \{X\}\}$ an X is a source set added to H' by step e2). Notice that the union of all C_i' in M' is a subset of K'. Since K' and $\underline{K}$ are equivalent, and all C_i' in M' are singletons, there exists an injection $f'' : M' \dashrightarrow \underline{C}$ such that for each C_i' in M' we have $|C_i'| \leq |f''(C_i')|$ and C_i' and $f''(C_i')$ are equivalent. Moreover, by Fact 3, for each C_i' in M', there exists no C_j' in C'' such that $f''(C_i') = f'(C_j')$. Hence, there exists an injection $f : C' \dashrightarrow \underline{C}$ such that $f(C'') = f'(C'')$ and $f(M') = f''(M')$. It follows that $|C'| \leq |\underline{C}|$ and, then, $|C'| = |\underline{C}|$ (i.e., f is a bijection) since $\underline{H}$ has the minimum number of source sets by assumption. Hence, also H' has the minimum number of source sets and, then, H' is a SMSE—hypergraph of H.

Let us now analyze the complexity of the algorithm. By Fact 2, steps a), c), d) and f) can be done in $O(|H|^2)$ time. By Proposition 1, step b) can be done in $O(s \cdot |H|)$ time, where s is the number of source sets of H. Step e) can be easily implemented in $O(s)$ time. Hence, the whole algorithm runs in $O(|H|^2)$ time, since $s < |H|$.

                                                                    Q.E.D.

Example 8. Let us consider the hypergraph **H** in Fig. 8a). A kernel of **H** is K ={{ C, D, E }, { B, G, H}}. The strong components of **H** are C_1 ={{A, B }, { C, D }, { C, D, E}}, C_2 ={{C}}, C_3 ={{D}}, C_4 ={{F}}, C_5 ={{G}}, C_6 ={{B, G, H}}. The source sets { C, D} and { C, D, E}are superfluous since the closure of **H** does not change if we replace the hyperarcs ({ C, D}, E) and ({ C, D, E}, F) by ({ A, B }, E) and ({ A, B}, F). Moreover, the hyperarc ({ B, G, H}, J) is redundant. Hence, by applying steps a), b), c) and d) of the algorithm for finding an SMSE–hypergraph of **H** we obtain the hypergraph **H"** in Fig. 8b). Note that **H"** is not strongly equivalent to **H**. Let us now perform step e) of the algorithm. We have that, during this step, the source set { C, D, E} in K is replaced by { A, B} and the hyperarc ({ B, G, H}, H) is added to **H**. Finally, by performing step f) the hyperarc ({ B, G, H}, H) is replaced by ({ B, G }, H) since the node **H** is redundant in { B, G, H }. The SMSE–hypergraph so obtained is shown in Fig. 8c).

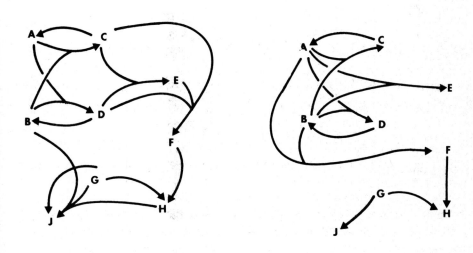

a) A hypergraph H.

b) A hypergraph H" (which is also an SME–hypergraph of H).

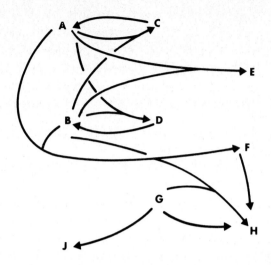

c) A hypergraph H' which is a minimal strongly equivalent
   hypergraph of H.

Figure 7.

5. Conclusions

In this paper the problem of determining minimal representations of equivalent directed hypergraphs has been considered. In particular the notion of kernel of a hypergraph has been introduced and a new concept of strong equivalence among hypergraphs has been studied, based on the equivalence of the corresponding kernels. The problem of determining a strongly equivalent hypergraph with the minimum number of source set has been analyzed. This problem, which has relevant applications in relational database research, has been shown to be solvable in polynomial time while other minimal representation problems for directed hypergraphs have been shown to be intractable in a previous paper [2]. Applications to various areas of computer science of this formalism and of the main results proved in this paper are being developped.

References

[1] Aho, A. V., M. R. Garey and J. D. Ullman, *"The Transitive Closure of a Directed Graph"*, SIAM Journal of Computing 1, (1972), 131-137.

[2] Ausiello, G., A. D'Atri and D. Saccà, *"Minimal Representations of Directed Hypergraphs"*, Rept. Istituto di Automatica della Università di Roma, R. 83-02, Submitted for publication, (Jan. 1983).

[3] Batini, C. and A. D'Atri, *"Database Design Using Refinement Rules"*, RAIRO Informatique Theorique 17:2, (1983), 97-119.

[4] Berge, C., *Graphs and Hypergraphs,* North Holland, Amsterdam, The Netherlands, (1973).

[5] Boley, H., *"Directed Recursive Labelnode Hypergraph: A New Representation Language"*, Artificial Intelligence 9, (1977), 49-85.

[6] Garey, M. R. and D. Johnson, *Computers and Intractability,* Freeman, San Francisco, California, (1979).

[7] Gnesi, S., U. Montanari and A. Martelli, *"Dinamic Programming as a Graph Searching: An Algebraic Approach"*, Journal of ACM 28:4, (1981), 737-751.

[8] Maier, D., *The Theory of Relational Databases,* Computer Science Press, Rockville, Maryland, (1983).

[9] Sacca', D., *"On the Recognition of Coverings of Acyclic Database Hypergraphs"*, Proceedings 2nd ACM SIGACT–SIGMOD Symposium on Principles of Database Systems, Atlanta, Georgia, (March 1983), 297-304.

[10] Ullman, J.D., *Principles of Database Systems,* second edition, Computer Science Press, Potomac, Maryland, (1982).

[11] Yannakakis M., *"A Theory of Self Locking Policies in Database Systems"*, Journal of ACM 29:3, (1982), 718-740.

Annals of Discrete Mathematics 25 (1985) 27-46

A LOCAL-RATIO THEOREM FOR APPROXIMATING THE WEIGHTED VERTEX COVER PROBLEM

R. BAR-YEHUDA and S. EVEN

Computer Science Department Technion - Istrael Institute of Technology
Haifa, Istrael 32000

A local-ratio theorem for approximating the weighted vertex cover problem is presented. It consists of reducing the weights of vertices in certain subgraphs and has the effect of local-approximation.

Putting together the Nemhauser-Trotter local optimization algorithm and the local-ratio theorem yields several new approximation techniques which improve known results from time complexity, simplicity and performance-ratio point of view.

The main approximation algorithm guarantees a ratio of $2 - \frac{1}{\kappa}$ where κ is the smallest integer s.t. $(2\kappa - 1)^{\kappa} \geq n$ (hence: ratio $\leq 2 - \frac{loglogn}{2logn}$) †

This is an improvement over the currently known ratios, especially for a "practical" number of vertices (e.g. for graphs which have less than 2400, 60000, 10^{12} vertices the ratio is bounded by 1.75, 1.8, 1.9 respectively).

1. Introduction

A *Vertex Cover* of a graph is a subset of vertices such that each edge has at least one endpoint in the subset. The *Weighted Vertex Cover Problem* (WVC) is defined as follows: Given a simple graph $G(V, E)$ and a weight function $\omega: V \to R^+$, find a cover of minimum total weight. WVC is known to be NP–Hard, even if all weigts are 1 [16] and the graph is planar [8]. Therefore, it is natural to look for efficient approximation algorithms.

Let A be an approximation algorithm. For graph G with weight functions ω, let C_A, C^* be the cover A produces and an optimum cover,

---

† *All log bases, in this paper, are 2.*

respectively. Define

$$R_A(G, \omega) = \frac{\omega(C_A)}{\omega(C^*)}$$ and let the performance ratio $r_A(n)$ be

$$r_A(n) = Sup\left\{R_A(G, \omega) \mid G = (V, E) \text{ where } n = |V|\right\}.$$

Many approximation algorithms with performance ratio ≤ 2 have been suggested; see, for example Table 1. No polynomial-time approximation algorithm A is known for which $r_A(n) \leq 2-\varepsilon$, where $\varepsilon > 0$ and fixed. Several approximation algorithms are known for which $R_A(G, \omega) \leq 2-\varepsilon(G)$, where ε depends on G; e.g. $\varepsilon(G) = \dfrac{2}{\Delta(G)}$

where $\Delta(G)$ is the maximum degree of the vertices of G.

In Section 2 we review the Nemhauser and Trotter [18] local optimization algorithm (NT) as well as the observation of Hochbaum [11] to use it for approximating WVC.

In Section 3, a new theorem, the 'Local-Ratio Theorem' is presented and proved. It consists of reducing the weights of vertices in certain subgraphs and has the effect of local approximation. As an example of its power we present a trivial proof for the correctness of a linear time approximation algorithm $COVER1$ with $r_{COVER1}(n) \leq 2$.

In Section 4, we present two approximation algorithms in which the NT algorithm and the local-ratio theorem are shown to be useful. The first algorithm $COVER2$ satisfies $r_{COVER2}(n) \leq 2-\dfrac{1}{\sqrt{n}}$ for general graphs, while for planar graphs $r_{COVER2}(n) \leq 1.5$ and its time complexity is $0(n^2 \log n)$. Hochbaum [12] obtained the same performance ratio, but we manage to avoid the time complexity of 4-coloring.

For our last algorithm, $COVER3$, we prove that $r_{COVER3}(n) \leq 1-\dfrac{1}{k}$, where k is the least integer s.t. $(2k-1)^k \geq n$. A similar result, but only for unweighted graphs, has been obtained independently by Monien and Speckenmeyer [17].

In Section 5, the Local-Ratio Theorem is extended to the *Weighted Set-Cover Problem.*

TABLE 1–SUMMARY OF APPROXIMATION RESULTS. (for NT, see Table 2)			
References	*Performance ratio* $\leq$	*Complexity for weighted*	*Complexity for unweighted*
[9]	2	not applicable	E
[11]	2	V^3	
[1]	2	E	
[12]	$2 - \dfrac{2}{\Delta}$	"NT"	"NT"
this paper	$2 - \dfrac{2}{\sqrt{V}}$	"NT"	"NT"
this paper	$2 - \dfrac{\log\log V}{2\log V}$	"NT"	EV
(For planar graphs)			
[12]	1.6	"NT"	$V^{1.5}$
[12]	1.5	"NT" + "4–COLORING"	"NT" + "4–COLORING"
this paper	1.5	"NT"	"NT"
[2]	$1\dfrac{2}{3}$	not applicable	V
[5]*	$1 + \varepsilon$	not applicable	Vlog V

* Although this result seems much better then the one shown 2 lines above for unweighted graphs, the algorithm is not useful for practical computations. For example for the algorithm to achieve a ratio ≤ 2 one may need ad many as $2^{2^{2^{160}}}$ vertices [6].For a similar algorithm see [2].

TABLE 2– "NT"'s Complexity (the same as MAX FLOW)		
$G = (V, E)$	*Weighted*	*Unweighted*
General	$E^{2/3}V^{5/3}$ or $EV \log V$	$E\sqrt{V}$
Planar	$V^2 \log V$	$V^{1.5}$

2. The Nemhauser and Trotter local optimization algorithm

In this short section we review the local optimization algorithm of Nemhauser and Trotter [18] which is very useful for approximations of WVC [12].

Let $G(V, E)$ be a simple graph. We denote by $G(U)$ the subgraph of G induced by $U \subseteq V$ and let $U' = \{u' | u \in U\}$. Define the weights of vertices in U' by $\omega(u') = \omega(u)$.

Algorithm NT

Input : $G(V, E), \omega$.

Phase 1:   Define a bipartite graph $B(V, V', E_B)$ where
$E_B = \{(x, y') | (x, y) \in E\}$.

Phase 2: $C_B \leftarrow C^*(B)$.

Output: $C_0 \leftarrow \{x | x \in C_B \ AND \ x' \in C_B\}$
$V_0 \leftarrow \{x | x \in C_B \ XOR \ x' \in C_B\}$

The following theorem states results of Nemhauser and Trotter.

The NT—Theorem: The sets C_0, V_0 which Algorithm NT produces, satisfie the following properties:

(i)   If a set $D \subseteq V_0$ covers $G(V_0)$ then $C = D \cup C_0$ covers G.

(ii)   There exists an optimum cover $C^*(G)$ such that $C^*(G) \supseteq C_0$.
   [(i) and (ii) are called the local optimally conditions.]

(iii) $\omega(C^*(G(V_0))) \geq 1/2 \, \omega(V_0)$.
   For a proof see [18]. An alternate proof is given in the APPEN-DIX.

The significance of the NT−Theorem, as pointed out by Hochbaum [12], is that the problem of approximating WVC can be limited to graphs $G(V_0)$ which satisfy (iii). A simple illustration of this approach was used by Hochbaum to obtain an algorithm with performace ratio ≤ 2 : Call $NT(G(V), \omega)$ to get C_0, V_0 and return $C = C_0 \cup V_0$.

Let us consider now the time-complexity of finding $C^*(B)$, which determines the time-complexity on NT.

For the unweighted case the problem can be converted into the maximum matching problem on B (see for example [3] which is of time-complexity $O(E\sqrt{V})$ (see [14]).

For the weighted case the problem can be converted into a maximum flow problem (see, for example [12]) which is of time-complexity $O(E^{2/3} V^{5/3})$ [10] or $O(EV \log V)$ [19]. For a summary of the results, see Table 2.

3. The local-ratio theorem

In a previous paper [2] we presented a local approximation technique for the vertex cover problem of unweighted graphs. In this section we present a local approximation technique for the vertex cover problem of weighted graphs. First, we present the following lemma:

Lemma: Let $G(V, E)$ be a graph and ω, ω_1 and ω_2 be weight functions on V, s.t. for every $v \in V$: $\omega(v) \geq \omega_1(v) + \omega_2(v)$. Let C^*, C_1^* and C_2^* be optimum covers of G with respect to ω, ω_1 and ω_2. It follows that: $\omega(C^*) \geq \omega_1(C^*) + \omega_2(C_2^*)$.

Proof: $\omega(C^*) = \sum_{v \in c^*} \omega(v)$

$$\geq \sum_{v \in c^*} (\omega_1(v) + \omega_2(v))$$

$$= \omega_1(C^*) + \omega_2(C^*)$$

$$\geq \omega_1(C^*) + \omega_2(C^*) \quad \text{[by the optimality of } C_1^* \text{ and } C_2^*]$$

$$\text{Q.E.D.}$$

Let $\bar{G}$ be an unweighted graph of $\bar{n}$ vertices, whose optimum cover contains $\bar{c}^*$ vertices. Define $\bar{r} = \bar{n}/\bar{c}^*$. Let A be an approximation algorithm for WVC and let $LOCAL_{\bar{G}}$ be the following algorithm:

Algorithm $LOCAL_{\bar{G}}$

Input: Graph $G(V, E)$, with weight function ω.

Phase 1:   Choose a subgraph $\tilde{G}(\tilde{V}, \tilde{E})$ of G which is isomorphic to $\bar{G}$.

Choose $0 \leq \delta \leq Min\{\omega(x) \mid x \in \tilde{V}\}$

Define $\omega_0(x) = \begin{cases} \omega(x) - \delta & \text{if } x \in \tilde{V} \\ \omega(x) & else. \end{cases}$

Phase 2:   Call $A(G, \omega_0)$ to get C_0.

Output: $C \leftarrow C_0$.

The Local-Ratio Theorem: $R_{LOCAL_{\bar{G}}}(G, \omega) \leq Max\{\bar{r}, R_A(G, \omega_0)\}$

Proof: Let c^* and c_0^* be the weights of the optimum covers of G with respect to ω and ω_0, and let $r = Max\{\bar{r}, R_A(G, \omega_0)\}$, then

$$\omega(C) \leq \omega_0(C) + \delta \cdot \bar{n} \quad \text{[by } |C \cap \tilde{V}| \leq \bar{n}]$$

$$\leq R_A(G, \omega_0) \cdot c_0{}^* + \bar{r} \cdot \delta \cdot \bar{c}^* \quad \text{[by definitions]}$$

$$\leq r \cdot (c_0{}^* + \delta \bar{c}^*) \quad \text{[by } r\text{'s definition]}$$

$$\leq r \cdot c^* \quad \text{[by the lemma]}$$

Q.E.D.

Let us consider now a corollary of the Local-Ratio Theorem. Let Γ be a finite family of graphs, and $r_\Gamma = Max\{\bar{r} \mid \bar{G} \in \Gamma\}$.

We denote by $G(U)$, $U \subseteq V$, the subgraph of $G(V, E)$ induced by U.

Algorithm $LOCAL_\Gamma$

Input: $G(V, E)$, ω.

Phase 0: **For** every $x \in V$ **do** $\omega_0(x) \leftarrow \omega(x)$ **end**

Phase 1: **For** every $\widetilde{G}(\widetilde{V}, \widetilde{E})$, subgraph of G which is isomorphic to some $\bar{G} \in \Gamma$, **do**

$$\delta \leftarrow Min\{\omega_0(x) \mid x \in \widetilde{V}\}.$$

For every $x \in \widetilde{V}$ **do** $\omega_0(x) \leftarrow \omega_0(x) - \delta$ **end**

end

Phase 2: $C_1 \leftarrow \{x \mid \omega_0(x) = 0\}$.

$V_1 \leftarrow V - C_1$.

Call $A(G(V_1), \omega_0)$ to get C_2.

Output: $C \leftarrow C_1 \cup C_2$.

The Local-Ratio Corollary:

$$R_{LOCAL_\Gamma}(G, \omega) \leq Max\{r_\Gamma, R_A(G(V_1), \omega_0)\}.$$

Proof: By induction on i, the number of iterations of Phase 1, during which $\delta > 0$.

For $i=0$ the claim is trivial.

Suppose the claim holds for i, and for some G, ω Phase 1 runs $i+1$ iterations during which $\delta > 0$. Let $\bar{G} \in \Gamma$ be the graph used in the first iteration during which $\delta > 0$. Observe that we can view the running of $LOCAL_\Gamma$ as an application of $LOCAL_{\bar{G}}$ where A (in Phase 2 of $LOCAL_{\bar{G}}$) is replaced by the remaining part of $LOCAL_\Gamma$. The inductive step is now an immediate consequence of the Local-Ratio Theorem.

Q.E.D.

In the applications of $LOCAL_\Gamma$ we shall refer to r_Γ as (a bound on) the local-ratio of Phase 1. Let us demonstrate a simple application of the Local-Ratio Corollary.

Algorithm *COVER*1

Input: $G(V, E)$, ω.

Phase 1:   **For** every $e \in E$ **do**

   Let $\delta = Min\{\omega(x)|x \in e\}$.

   For every $x \in e$ **do** $\omega(x) = \omega(x)-\delta$ **end**

   end

Output: $C \leftarrow \{x \mid \omega(x) = 0\}$.

This approximation algorithm is essentially the one we described in [1][1].

Proposition: For algorithm *COVER*1.

(1)   The time complexity is $O(E)$

(2)   Its performance ratio ≤ 2

Proof:

(1)   The number of operations spent on each edge is bounded by a

---

[1] *In* [1] *we used a global rather then a local point of view.*

constant.

(2) Using the Local-Ratio Corollary, with Γ which is a single edge.

Q.E.D.

4. Putting together NT and the local-ratio theorem

Hochbaum [12] suggested the following approach to approximate WVC: Let $G(V, E)$, ω be the problem's input, such that $\omega(C^*(G)) \geq 1/2\ \omega(V)$. (This is achieved by the NT algorithm). Color G by k colors and let I be the "heaviest" monochromatic set of vertices. The cover produced is $C = V - I$. It follows that

$$\frac{\omega(C)}{\omega(C^*)} = \frac{\omega(V) - \omega(I)}{\omega(C^*)} \leq \frac{\omega(V) - \omega(V)/\kappa}{1/2\ \omega(V)} = 2 - \frac{2}{k}.$$

For general graphs she gets the ratio $2 - \dfrac{2}{\Delta}$ (Δ is the maximum degree) and for planar graphs ($k = 4$) the performance ration ≤ 1.5 in time-complexity of NT and 4-coloring.

We suggest the use of a preparatory algorithm in which all triangles are omitted (with local-ratio 1.5) and therefore, the residual graph is easier to color.

Algorithm *COVER 2*

Input: $G(V, E)$, ω, k.

Phase 1:   [Triangle elimination]

For every triangle $T\ (T \subseteq V)$ **do**

   Find $\delta = Min\ \{\omega(x) \mid x \in T\}$

   For every $x \in T$ **do** $\omega(x) \leftarrow \omega(x) - \delta$ **end**

end

$C_1 \leftarrow \{x \mid \omega(x) = 0\}$

$$V_1 \leftarrow V - C_1.$$

Phase 2:   Call $NT(G(V_1), \omega)$ to get C_0, V_0.

Phase 3:   Find a cover approximation, C_2, of $G(V_0)$,

                 by using k-coloring (as in Hochbaum's approach).

Output: $C \leftarrow C_1 \cup C_0 \cup C_2$.

For $k \geq 4$, $R_{COVER2}(G, \omega) \leq 2 - \dfrac{2}{k}$, since the local-ratio of Phase 1 is 1.5 (here Γ contains of a single graph $\bar{G}$ which is a triangle), the local-ratio of Phase 2 is 1 and the local-ratio of Phase 3 is $2 - \dfrac{2}{k}$.

Since the graph colored in Phase 3 is triangle-free, we get the following additional results:

(1)   For general graphs $r_{COVER2}(n) \leq 2 - \dfrac{1}{\sqrt{n}}$ by using Wigderson's approach [20] for coloring a triangle-free graph by $k = 2\sqrt{n}$ colors in linear time [21].

(2)   For planar graphs $r_{COVER2}(n) \leq 1.5$, and the time-complexity of 4-Coloring a triangle-free planar graph is linear. (One uses the fact that in such graphs there is always a vertex of degree≤ 3. See, for example, [13]. This prevents the need to use a more complex 4-Coloring algorithms.

Note that the 1.5 performance ratio, for unweighted planar graphs, can be achieved by the "1 + ε algorithm" of [5] (or [2]), however their algorithm is expo-exponential time w.r.t. $1/\varepsilon$ and for $\varepsilon = 0.5$ is not practical.

Before we present our main algorithm we need a few preliminaries. Let the triple $(G(V, E), \omega, k)$ (where G is a graph with weight function ω and k is a positive integer) be called *proper* if the following conditions hold:

(i) $(2k-1)^k \geq |V|$.

(ii) There are no odd circuits of length $\leq 2k - 1$.

(iii) $\omega(C^*) \geq 1/2 \, \omega(V)$.

In the following procedure the statements in square brackets are added for the analysis only, and variables j (integer), C_0, V_0, C_1, V_1,.. (sets), are used only in the brackets.

Procedure *COVER.PROPER*

Input: proper $(G(V, E), \omega, k)$.

Phase 0: V' $V' \leftarrow V$, $C' \leftarrow \phi$ · $[j \leftarrow 0]$

Phase 1:   **While** $V' \neq \phi$ **do**

Find $v \in V'$ s.t. $\omega(v) = Max\{\omega(u) \mid u \in V'\}$.

Let $A_0, A_1, ..., A_k$ be the first $k + 1$ layers†

of a Breadth-First-Search (BFS‡) on $G(V')$ starting with

$A_0 = \{v\}$

Define $B_{2t} = \bigcup_{i=0}^{t} A_{2i}$ and $B_{2t+1} = \bigcup_{i=0}^{t} A_{2i+1}$

(for $t=0, 1, 2, ...$)

$f \leftarrow Min\{s \mid \omega (B_s) \leq (2\kappa-1) \cdot \omega(B_{s-1})\}$.

Add B_f to C'. $[C_j \leftarrow B_f]$

Remove $B_f \cup B_{f-1}$ from V'. $[V_j \leftarrow B_f \cup B_{f-1}, j \rightarrow j+1]$

end

Output: $C \leftarrow C'$.

Proposition 1: Procedure *COVER.PROPER* satisfies the following properties:

(1)   In every application of Phase 1, $f \leq \kappa$.

(2)   In every application of Phase 1, B_{f-1} is an independent set in $G(V')$.

(3) C covers G.

---

† *Starting with some* m, A_m, A_{m+1}, ..., A_κ *may be empty.*

‡ *See for example* [7].

(4)   For every iteration j, $\omega(C_j) \leq (1 - \dfrac{1}{2k}) \cdot \omega(V_j)$.

(5) $\omega(C) \leq (1 - \dfrac{1}{2k}) \cdot \omega(V)$.

(6) $R_{COVER.PROPER}(G) \leq 2 - \dfrac{1}{k}$.

(7)   The time complexity is $0(|V| \log |V| + |E|)$.

Proof:

(1)   Assume the contrary.
      Thus, for every $s \leq k$, $\omega(B_s) > (2k-1) \cdot \omega(B_{s-1})$. Thus,

$$\omega(B_k) > (2k-1)^k \cdot \omega(B_0) \quad \text{[by the assumption]}$$

$$\geq |V| \cdot \omega(B_0) \quad \text{[by (i) of the definition of proper]}$$

$$= |V| \cdot \omega(v) \quad [B_0 = A_0 = \{v\}]$$

$$\geq |V'| \cdot \omega(v) \quad [V' \subseteq V]$$

$$\geq \omega(V') \quad \text{[by } v\text{ 's definition]}$$

$$\geq \omega(B_k) \quad [B_k \subseteq V']$$

which is absurd.

(2)   An edge between two vertices of A_s, $s < k$, implies the existence
      of an odd-circuit of length $\leq 2k-1$. This contradicts condition
      (ii) of properness.

(3)   For every iteration j, all edges which are (indirectly) deleted are
      covered by the current C_j, which joins C'. This follows from the
      properties of BFS and (2) above.

(4)   For every iteration j, $\omega(B_f) \leq (2k-1) \cdot \omega(B_{f-1})$. Since $B_f = C_j$,
      and $B_{f-1} = V_j - C_j$, we have $\omega(C_j) \leq (2k-1) \cdot [\omega(V_j) - \omega(C_j)]$.
      This implies the stated inequality.

(5)   By summation of the inequality of (4) for every j.

(6) By definition $R_{COVER.PROPER}(G, \omega) = \dfrac{\omega(C)}{\omega(C^*)}$. Property (5) above and condition (iii) of properness imply that

$$R_{COVER.PROPER}(G, \omega) \leq \frac{(1-\dfrac{1}{2k}) \cdot \omega(V)}{1/2\,\omega(V)}.$$

(7) We may start the algorithm by sorting the vertices according to their weights, which requires $0(|V|log|V|)$ steps. This complexity includes, now, the total time used in Phase 1 for finding $Max\left\{\omega(u) \mid u \in V'\right\}$.
It is not necessary to continue the BFS of Phase 1, beyond layer f. Thus, each such search is linear in the number of edges to be eliminated from $G(V')$, and the total time spent in the search of Phase 1 is linear in $|E| + |V|$.
Thus (7) follows.

Q.E.D.

Now, the main algorithm.

Algorithm *COVER3*

Input: $G(V,E), \omega$

Phase 0: Find the least integer k s.t. $(2k-1)^k \geq |V|$.

Phase 1: [Elimination of short odd circuits with local-ratio $\leq 2-\dfrac{1}{k}$].

For every odd circuit $D \subseteq V$ s.t $|D| \leq 2k-1$ do

$\quad \delta \leftarrow Min\left\{\omega(x) \mid x \in D\right\}$
$\quad$ For every $x \in D$ do $\omega(x) \leftarrow \omega(x)-\delta$ end

end
$C_1 \leftarrow \left\{x \mid \omega(x) = 0\right\}.$
$V_1 \leftarrow V-C_1.$

Phase 2: Call $NT(G(V_1), \omega)$ to get $C_0, V_0.$

Phase 3:   Call $COVER.PROPER(G(V_0), \omega, k)$ to get C_2.

Output: $C \leftarrow C_1 \cup C_0 \cup C_2$.

Proposition 2:   Algorithm $COVER3$ satisfies the following properties:

(1) $r_{COVER3}(n) \le 2 - \dfrac{1}{k}$.

(2)   Its time complexity is the same as NT's. (see Table 2)

(3)   for unweighted graphs its time complexity is $O(|V| \cdot |E|)$.

Proof:

(1)   The combination of Phase 2 and 3 yields an algorithm with performance ratio $\le 2 - \dfrac{1}{k}$, since the NT algorithm performs local-optimization (ratio=1) and, for proper graphs, $COVER.PROPER$ has performance ratio $\le 2 - \dfrac{1}{k}$ [by Proposition 1(6)]. Let $\bar{G}_l$ be a simple circuit of length $2l-1$ thus, $\bar{n}_l = 2l-1$ and $\bar{c}_l^* = l$; therefore, $\bar{r}_l = (2l-1)/l = 2 - 1/l$. Consider, now, the Local-Ratio-Corollary where, $\Gamma = \{\bar{G}_l \mid l \le k\}$, thus, $r_\Gamma = Max\{\bar{r}_l \mid l \le k\} = 2 - 1/k$ and (1) follows.

(2)   Let us perform Phase 1 in a slightly different way: Choose a vertex v, and build the first k layers of the BFS starting from v. If there is an edge $u-\omega$, where u and ω belong to the same layer, then an odd-length-circuit D has been detected (see for example [15]). In this case we find $\delta = Min\{\omega(x) \mid x \in D\}$, reduce the weights of the vertices in D by δ and the vertices whose weight is zero are eliminated from the representation of the graph (for the purpose of performing Phase 1). If no such edge (closing an odd circuit) is detected then v is eliminated from the representation of the graph, since no odd-circuit of length $\le 2k-1$ passes through it. In any case at least one vertex is eliminated for each BFS. Thus the time complexity is $O(|V| \cdot |E|)$.
In Phase 2, the best known time-complexity of the NT algorithm

(table 2) is greater then $|V| \cdot |E|$. Phase 3 requires $O(|E|+|V|log|V|)$, by Proposition 1(7).

Thus, the whole algorithm is of time complexity as the NT algorithm.

(3) For unweighted graphs, the NT algorithm can be performed in $O(|E|\sqrt{|V|})$ time and therefore, the algorithm is of time complexity $O(|V| \cdot |E|)$.

Q.E.D.

Corollary: $r_{COVER3}(n) < 2 - \dfrac{loglogn}{2logn}$.

Proof: Define $g(n) = \dfrac{logn}{loglogn}$ which is monotone increasing for $n \geq 16$.

By k-th definition $(2k - 3)^{k-1} < n$. Thus, $g((2k-3)^{k-1}) < g(n)$.

We want to show that $k < 2g(n)$. Thus, it suffices to show that $k < 2g((2k - 3)^{k-1})$. This is an exercise in elementary mathematics.

Q.E.D.

5. Extending the local ratio theorem for the weighted set cover problem

Let HWVC be the following problem. Given a hypergraph $G=(V,E)$ with weight function $\omega: V \to R^+$, find a set $C \subseteq V$ of minimum total weight s.t. for every $e \in E$, $|e \cap C| \geq 1$. The local-Ratio Theorem and its Corollary hold also for HWVC. Algorithm $COVER1$ can be applied directly to HWVC with $r_{COVER1} \leq \Delta_E$ [where Δ_E is the maximum edge-degree (or cardinality) in G]. Its running time is linear in the length of the problem's input ($\Sigma |e|$). HWVC is actually the Weighted-Set-Cover Problem[2] and is an extension of WVC [$\Delta_E = 2$]. Even

[2]Chvatal [4] gets, performance-ratio $\leq \sum_{j=1}^{\Delta} \dfrac{1}{j} = 0(log \Delta)$.

[where Δ is the maximum vertex degree in G].

though we get performance-ratio$\leq \Delta_E$ in linear time, we suspect that for any fixed Δ_E, there is no polynomial time approximation algorithm with a better constant performance ratio, unless P=NP [even for the unweighted case], (this is an extension of a conjecture of Hochbaum).

6. Appendix – NT theorem

Let $G(V, E)$ be a simple graph. We denote by $G(U)$ the subgraph of G induced by $U \subseteq V$ and let $U' = \{u' \mid u \in U\}$. Define the weights of vertices in U' by $\omega(u') = \omega(u)$.

Nemhauser and Trotter [18] presented the following local optimization algorithm:

Algorithm NT

Input: $G(V, E), \omega$.

Phase 1:   Define a bipartite graph $B(V, V', E_B)$ where
$E_B = \{(x, y') \mid (x, y) \in E\}$.

Phase 2: $C_B \leftarrow C^*(B)$.

Output: $C_0 \leftarrow \{x \mid x \in C_B \text{ AND } x' \in C_B\}$
$V_0 \leftarrow \{x \mid x \in C_B \text{ XOR } x' \in C_B\}$

The following theorem states results of Nemhauser and Trotter, but our proof is shorter and does not use linear programming arguments.

The NT–Theorem: The sets C_0, V_0 which Algorithm *NT* produces, satisfies the following properties:

(i)   If a set $D \subseteq V_0$ covers $G(V_0)$ then $C = D \cup C_0$ covers G.

(ii)   There exists an optimum cover $C^*(G)$ such that $C^*(G) \supseteq C_0$.

(iii) $\omega(C^*(G(V_0))) \geq 1/2 \, \omega(V_0)$.

Proof: Define $I_0 = \{x \mid x \notin C_B \text{ AND } x' \notin C_B\} = V-(V_0 \cup C_0)$

Let $(x, y) \in E$. In order to prove (i) we need to show that either $x \in C$ or $y \in C$.

Case 1: $x \in I_0$, i.e. $x, x' \notin C_B$. Thus, $y, y' \in C_0$ and therefore $y \in C_0$.

Case 2: $y \in I_0$. Same as Case 1.

Case 3: $x \in C_0$ or $y \in C_0$. This case is trivial.

Case 4: $x, y \in V_0$. Thus, either $x \in D$ or $y \in D$.

In order to prove (ii), let $S = C^*(G)$. Define
$S_V = S \cap V_0$, $S_C = S \cap C_0$, $S_I = S \cap I_0$ and $\bar{S}_I = I_0 - S_I$.
Let us show that:

$$C_{B_1} = (V-\bar{S}_I) \cup S'_c \text{ covers } B. \tag{*}$$

Let $(x, y') \in E_B$. We need to show that either $x \in C_{B_1}$ or $y' \in C_{B_1}$.

Case 1: $x \notin \bar{S}_I$. Thus, $x \in V-\bar{S}_I [\subseteq C_{B_1}]$

Case 2: $x \in \bar{S}_I$. Thus, $x \in I_0$, $x \notin S$ and therefore $x \notin C_0$. It follows that $y \in S$ [since S covers (x, y)] and $y \in C_0$ [by Case 1, in the proof of (i)].
Thus, $y \in S \cap C_0 [=S_c]$ and therefore $y' \in S'_c [\subseteq C_{B_1}]$.

This proves (*). Now,

$$\begin{aligned}
\omega (V_0) + 2\omega(C_0) &= \omega(V_0 \cup C_0 \cup C'_0) \\
&= \omega(C_B) && \text{[by definitions of } V_0, C_0] \\
&\leq \omega (C_{B_1}) && \text{[by (*) and optimality of } C_B] \\
&= \omega ((V-\bar{S}_I) \cup S'c) \\
&= \omega(V_0 \cup C_0 \cup S_I \cup S'_c) \\
&= \omega(V_0) + \omega(C_0) + \omega (S_I) + (S_c).
\end{aligned}$$

It follows that $\omega(C_0) \leq \omega (S-S_V)$. Thus, $\omega(C_0 \cup S_V) \leq \omega(S)$. However, $C_0 \cup S_V$ covers G [by (i)] and therefore $C_0 \cup S_V$ is an optimum cover of G and contains C_0. This proves condition (ii).

In order to prove (iii), assume S_0 is an optimum cover of $G(V_0)$. By (i), $C_0 \cup S_0$ covers G, and by B's definition $C_0 \cup C'_0 \cup S_0 \cup S'_0$ covers B. Thus,

$$\omega(V_0) + 2\omega(C_0) = \omega(C_B)$$

$$\leq \omega(C_0 \cup C'_0 \cup S_0 \cup S'_0)$$
$$\text{[by optimality of } C_B]$$
$$= 2\omega(C_0) + 2\omega(S_0).$$

Therefore, $\omega(V_0) \leq 2\omega(S_0)$.

Q.E.D.

References

[1]   Bar-Yehuda, R., and S. Even, "A Linear Time Approximation Algorithm for the Weighted Vertex Cover Problem", *J. of Algorithms,* vol. 2, 1981, 198-203.

[2]   Bar-Yehuda, R., and S. Even "On Approximating a Vertex Cover for Planar Graphs", *Proc. 14th Ann. ACM Symp. Th. Computing,* 1982, 303-309.

[3]   Bondy, J. A., and U.S.R. Murty, *Graph Theory and Applications,* North Holland, 1976. Theorem 5.3, page 74.

[4]   Chvatal, V., "A Greedy-Heuristic for the Set-Covering Problem", *Math. Operations Res.,* vol. 4, no. 3, 1979, 233-235.

[5]   Chiba, S., T. Nishizeki, and N., Saito, "Applications of the Lipton and Tarjan's Planar Separator Theorem", *J Information Proc.,* 4, 1981, 203-207.

[6]   Djidjev H. N.   "On the Problem of Partitioning Planar graphs", *SIAM J. on Alg. and Disc. Meth.* 3, 2, 1982, 229-240.

[7]   Even, S., *Graph Algorithms,* Comp. Sci. Press, 1979.

[8]   Garey, M. R., and D. S. Johnson, *Computers an Intractability; A Guide to the Theory of NP-Completeness;* Freeman, 1979.

[9]   Gavril, F. see [8] page 134.

[10] Galil, Z., "A new Algorithm for the Maximal Flow Problem", *Proc. 9th Ann. Symp. on Foundations of Computers Science,* 1978, 231-245.

[12] Hochbaum, D. S. "Approximation Algorithms for the Set Covering and Vertex Cover Problems, *"SIAM J. Comput.,* vol. 11, 3, 1982, 555-556.

[12] Hochbaum, D. S., "Efficient Bounds for the Stable Set, Vertex Cover and Set Packing Problems", *Dis. App. Math.* vol. 6, 1983, 243-254.

[13] Harary, F., *Graph theory,* Addision-Wesly, 1972 (revised).

[14] Hopcroft, J. E., and R. M. Karp, "An $n^{5/2}$ Algorithm for Maximum Matching in Bipartite Graphs," *SIAM J. Comput.,* 2, 1973, 225-231.

[15] Itai, A., and M. Rodeh, "Finding a Minimum Circuit in a Graph", *SIAM J. Comput.* vol. 7, 4, 1978, 413-423.

[16] Karp. R. M. "Reducibility among Combinatorial Problems", in R. E. Miller and J. W. Thatcher (eds.), *Complexity of Computer Computations,* Plenum Press, New York 85-103.

[17] Monien, B. and E. Speckenmeyer, "Some further Approximation Algorithms for the Vertex Cover Problem", to be presented in *the 8th Coll. on Trees in Algebra and Programming.* Univ. degli Studi de L'Aquila, Italy. March 1983.

[18] Nemhauser, G. L. and L. E. Trotter, Jr., "Vertex Packing: Structural Properties and Algorithms" *Mathematical Programming,* vol. 8, 1975, 232-248.

[19] Sleator, D. D. K., "An *0(mnlogn)* Algorithm for the Maximum Network Flow", Doctoral Dissertation, Dept. Computer Science, Stanford University, 1980.

[20] Wigderson, A., "Improving the Performance Guarantee for Approximate Graph Coloring", *Journal of Association for Computing Mechinery,* vol. 30, 4, 1983, 729-735.

[21] Wigderson, A., private communication.

Annals of Discrete Mathematics 25 (1985) 47-64

DYNAMIC PROGRAMMING PARALLEL PROCEDURES FOR SIMD ARCHITECTURES

Paola BERTOLAZZI

Istituto di Analisi dei Sistemi ed Informatica
C.N.R., Viale Manzoni 30, Roma

Dynamic Programming is a general technique for solving optimization problems such as linear and not linear programming, graph problems, image processing problems. Algorithms based on dynamic programming principles require very high amount of computation time: hence it seems to be necessary to resort on parallel computing in order to decrease the amount of computation.

Previous proposals of parallel procedures for Dynamic Programming were made both for Single Instruction Multiple Data stream (SIMD) architectures with centralized control and Multiple Instruction Multiple Data (MIMD) architectures.

In this paper we give two parallel procedures for SIMD architectures with synchronous and decentralized control. In these procedures the communication is performed in such a way that the communication phase length is kept constant during the entire run: this fact allows the decentralized control. We show that using a communication procedure due to Nassimi and Sahni we can solve the most general dynamic programming problem, while a more simple communication procedure is sufficient to solve dynamic programming problems whose state transition function is invertible.

1. Introduction

Dynamic programming is a powerful technique for solving large classes of optimization problems, such as integer programming, graph problems (shortest path, graph partitioning), pattern recognition [1], [2], [3].

In general, however, the method requires a great amount of computation, which makes it unsuitable especially in real time applications (i.e. image processing problems [13]).

With the advent of parallel architectures many efforts have been devoted to the design of parallel algorithms for many classes of problems; the goal was always to reduce the best serial computation time of these problems by a factor which is a function of the number of available processors [4], [5].

If T is the time of the best serial algorithm for a given problem, the best parallel time obtainable for this problem on a parallel machine with P processors is $0(T/P)$, where P is called the optimal speed-up ratio, unless a parallel procedure is derived which leads to a jet better sequential procedure.

In general the optimal speed-up ratio is not reached because of the overhead for carrying out the interprocessor communication [5].

As far as dynamic programming is concerned Larson proposed two parallel procedures for a general dynamic programming problem on a very theoretical computation model [6]. These two procedures are then evaluated on a SIMD architecture with centralized control (a master-slaves architecture with P processors) by Al Dabass in [7]. In this implementation the communication is carried out by the master processor, which creates a bottleneck in the system: all the communication among the P processors are performed serially; consequently a term $0(P)$ must be added to the expression of the parallel time $0(T/P)$ and a near optimal speed-up ratio can be reached only for a very small P (see [14] for this analysis).

The problem of communication is studied by Guibas et alii in [8], where they propose a parallel dynamic programming procedure for a particular problem: the optimal parenthesization. The parallel procedure is implemented on a mesh connected computer. The particular structure of the state transition function allows to pipeline data through the interconnnections; the technique of pipelining results in a maximum exploitation of the parallelism both in the computation and in the transmission. For this well structured problem the algorithm reaches the optimal speed-up ratio.

Finally we want to quote the proposal due to Bertsekas [9] of a distributed algorithm for dynamic programming on a MIMD architecture.

In this paper we give two parallel procedures for dynamic programming, both studied for two different SIMD architectures: the mesh connected computer and the orthogonal trees. The two procedures make use of suitable communication procedures which warrantee that the length of the communication phase is independent on the input data: this fact makes it possible the distributed control. The first procedure can solve any dinamic programming problem; for this procedure we show that augmenting the degree of parallelism in computation and in communication does not decrease significantly the parallel time.

The second procedure is good for solving dynamic programming problems which satisfy the assumption of invertibility of the state transition function; for this second procedure we show that the solution with less processors and a lower degree of parallelism in communication reaches the optimal speed-up ratio; on the other hand the use of a model with a greater number of processors and more parallelism in communication brings to a reduction of the parallel time from L to logL, although the optimal speed-up ratio is no more obtained.

In section 2 we give the definition of dynamic programming problem. In section 3 we describe the two parallel procedures. In section 4 we compare the two procedures and give an evaluation of them in terms of the AT^2 measure, in the case one would implement them using the VLSI technology.

2. Dynamic programming.

Dynamic Programming (D.P.) is a methodology for representing problems involving the optimization of a sequence of decisions (multi-stage decision problems) [1] [2] [3].

To introduce the D.P. ideas, consider the problem of minimizing the separable cost function:

$$z = g(s, u, j) \qquad\qquad 2.1$$

where the following system equation:

$$s = f(s, u, j) \qquad j = 0, 1 \ldots . N\text{-}1 \qquad\qquad 2.2$$

and the constraints:

$$s \in S \subset R^n$$

$$u \in U \subset R^m$$

must be satisfied.

Dynamic programming formulation

The above multi-stage decision problem can be solved by decomposing it into a sequence of subproblems such that:

a)   one subproblem has a relatively simple solution

b)   relations may be obtained linking together the subproblems.

The well known Bellman's principle of optimality suggests the following formulation for the above problem:

$$\text{find } z^* = V(s, 0)$$

where the following recursion equation:

$$V(s, j) = \min_{u \in U}(g(s, u, j) + V(f(s, u, j), j + 1)) \qquad\qquad 2.3$$

$$V(s, N + 1) = \infty$$

must be satisfied.

Dynamic programming computational procedure

The most general procedure to solve a D.P. problem is the so called iterative functional procedure (2). In this method the set of admissible decisions U and the set of admissible states S are quantized into a finite number of values. The set U then consists of the elements

$$U = u_1, u_2, \ldots u_M$$

and the set S of the elements

$$S = s_1, s_2, \ldots s_L$$

Notation: let us call

$u_{e/1}$ the value of the decision variable

s_h the value of the present state

s_k the value of the next state

The iterative functional procedure is given here in natural language:

for each stage j,   j = N, ... 1
    for each state s_h
        for each decision $u_{k/1}$

        compute g (s_h, u_1, j)
        compute $s_k = f(s_h, u_1, j)$
        compute V $(s_h, j) = \min(V(s_h, j)$ and
            g $(s_h, u_1, j) + V(s_k, j+1))$

Let t_b be the time needed to execute the instructions of the inner loop (basic evaluation).

The computation time can be expressed in terms of the number of basic evaluations times the time t_b to perform such an evaluation; the number of evaluations of

$$g(s_h, u_1, j) + V(f(s_h, u_1, j), j+1)$$

is once per quantized decision for each quantized state for each stage; the total time will be:

$$T = M*L*N*t_b$$

This value is very large in most of the practical applications. In fact as s and u are vectors with n and m components, L and M grow exponentially with n and m.

This fact makes D.P. unactractive both for very large applications and for those problems that require the solution is given in real time (i.e. image processing [13]). A possible solution to this problem is represented by parallel processing.

Of course parallel processing can not reduce the asymptotical complexity in the case of exponential grow, unless an exponential number of processors is used. However the computation time can be reduced by a factor which is proportional to the number of processors adopted, making parallel processing actractive in most practical applications.

3. The parallel procedures.

Parallel computation models

Our parallel procedures are studied on two parallel computation models: the mesh connected computer and the orthogonal trees. The mesh can be formally described as follows: the P processing elements, indexed 0 through P−1 are logically arranged as in a q-dimensional array of size $p_1 * p_2 * \ldots * p_q = P$. The processor at location $i_1, \ldots i_j, \ldots i_q$ is connected to the processors at location $i_1, \ldots i_j + 1, \ldots i_q$ $0 < j \leqslant q$. The first procedure is described with reference to the 2-dimensional mesh, while the second procedure is thought for a 1-dimensional array; both these structures are very realistic (see ILLIAC IV, DAP, and the Programmable Systolic Array of the Carnegie Mellon University).

The orthogonal trees is constructed starting from a $p_1 * p_2$ matrix of nodes (with p_1 and p_2 power of two). Adding nodes where they are needed, a complete binary tree is built for each row (column) of the matrix, having the nodes of such a row (column) as leaves. We will refer to such trees as to row (column) trees.

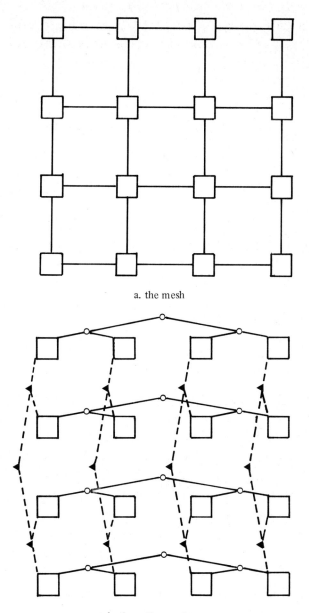

a. the mesh

b. the orthogonal trees

Fig. 1

An example of a mesh and an orthogonal trees for $p_1 = 4$ and $p_2 = 4$ is given in fig. 1. In the following we assume that our architectures have an unlimited number of processors to simplify the description of the two procedures; the results in the case of limited number of processors, easily deducible from the results in the case of unlimited number, are also given.

Decomposition into modules

We give the decomposition into modules in the case of unlimited parallelism, that is in the mesh with $P = L$ processors and in an orthogonal trees with $L * L$ or $L * M$ processors.

In the case of the mesh, in both the procedures, the number of processors equals the number of the state values. Each processor h of the mesh is dedicated to the computation concerning the status s_h. It stores only the information concerning the status s_h, that is the value of the status and the value of $V(s_h, j)$, for all j's. The module each processor executes is the set of instructions in the "for each state" loop of the serial procedure. This form of decomposition is called the "parallel states method" [6].

In the case of the orthogonal trees each row h of the matrix is devoted to the computation concerning the status s_h: in this row each processor is devoted to compute the quantity:

$$g(s_h, u_l, j) + V(f(s_h, u_l, j), j+1)$$

concerning the value of u associated to it. We call this kind of decomposition the "parallel states and parallel decisions method".

Procedure P1

The h-th processor of the $L * L$ mesh computes, for each stage, the value of $V(s_h, j)$ concerning the state s_h; it executes the following sequence of instructions:

0. $V(s_h, j) = \infty$

for each $u_l \in U$ pardo

1. compute $g(s_h, u_l, j)$

2. compute $s_k = f(s_h, u_l, j)$

3. read $V(s_k, j+1)$ stored in processor k

4. compute $V(s_h, j) = \min (V(s_h, j) \text{ and } g(s_h, u_l, j) + V(s_k, j+1))$

endfor

Each processor executes step 1 to 4 M times for each stage, that is N*M times in total.

The module executed by each processor differs from the sequential algorithm only in step 3, which represents the communication phase.

Thus the total time to execute one module is given by $t_b + t_c$ where t_c is the time to perform the communication phase.

The communication phase can be realized using a procedure of Nassimi and Sahni [11], based on the parallel sort algorithm of Kung [12]: it requires a time $O(L)$.

The procedure of Nassimi requires the same time also if there are undefined values of the next state and if more than one request reaches the same processor. In fig. 2 the steps 1 to 4 of procedure P1 are shown.

The above communication procedure works also in the case of limited number of processors, that is when $P < L$. In this case the time needed to execute the module is $O(L/P * (t_b + \sqrt{P}))$. Also in this case it works also if there are undefined values of the next state and if more than L/P requests arrive to the same processor.

This implies that the parallel procedure P1 can solve the most general dynamic programming problem in a sinchronous system with distributed control.

We now show that if we use more parallelism we do not obtain a large decrease of the time. Our reasoning is developed in the case of unbounded parallelism.

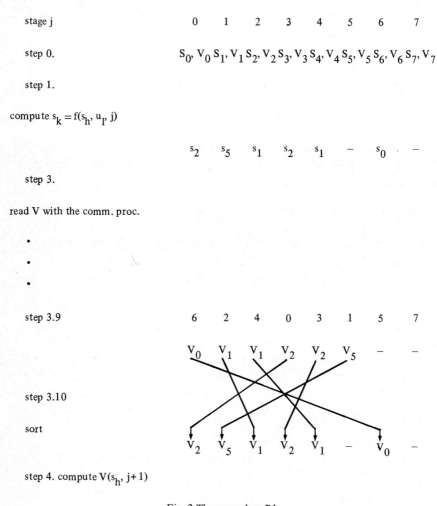

stage j   0   1   2   3   4   5   6   7

step 0. $S_0, V_0 \, S_1, V_1 \, S_2, V_2 \, S_3, V_3 \, S_4, V_4 \, S_5, V_5 \, S_6, V_6 \, S_7, V_7$

step 1.

compute $s_k = f(s_h, u_l, j)$

         s_2 s_5 s_1 s_2 s_1 — s_0 —

step 3.

read V with the comm. proc.

·

·

·

step 3.9   6   2   4   0   3   1   5   7

         V_0 V_1 V_1 V_2 V_2 V_5 —   —

step 3.10

sort

         V_2 V_5 V_1 V_2 V_1 — V_0 —

step 4. compute $V(s_h, j+1)$

Fig. 2 The procedure P1

Let us assume to have an orthogonal trees architecture with L∗M processors: in this architecture row h is devoted to computations concerning the state s_h and column 1 is devoted to computations concerning the decision u_1. The parallel algorithm in this architecture is the following:

step 1. Each processor h compute the value $s_k = f(s_h, u_1, j)$

step 2. A communication procedure, performed on each column tree, brings the information $V(s_k, j+1)$ in the appropriate processor

step 3. Each processor h computes $g(s_h, u_1, j) + V(s_k, j+1)$

step 4. The minimum of these values is computed along each row of the mesh and stored in all the processors of the row.

Step 2 of the algorithm requires a time that is $O(L)$: the communication procedure in fact will be based on a sort algorithm, which requires $O(L)$ time in a tree.

Step 4 requires $O(logM)$ in the orthogonal trees.

The total time in the orthogonal trees is $O(N * L)$, which is better than the time in the $\sqrt{L} * \sqrt{L}$ mesh only if $M > \sqrt{L}$.

Procedure P2

Let us now assume that the state transition function satisfies the following hypothesis:

hp. $\forall j = N, \ldots 1 , \forall s_h, s_k \in S * S$

there exists a unique $u_1 \in U$ such that $s_k = f(s_h, u_1, j)$

that is f is invertible with respect to u

In this case

$$V(s_h, j) = \min_{s_k \in S}(g(s_h, f^{-1}(s_h, s_k), j) + V(s_k, j+1)) \qquad 3.1$$

We first give the procedure on an one dimensional mesh with L processors.

Then we show that it can be implemented on an orthogonal trees architecture with L * L processors in time that is significantly lower than the time for the mesh.

To compute expression 3.1 processor h of the mesh must read the pair of values s_k, $V(s_k, j+1)$ associated to the processors $k \neq h$.

The transmission of these values can be carried out by simply shifting the data one step on the right (left) before the computation phase.

For each stage j = N, . . . 1 the h-th processor computes $V(s_h, j)$ executing the following set of instructions:

0. $V(s_h, j) = \infty$; $S1 = s_h$; $V1 = V(s_h, j+1)$

for i = 1 to L pardo

    1. read S1 and V1 stored in the processor number
      (h+1) mod L (shift)

    2. compute $u_1 = f^{-1}(s_h, S1)$

    3. compute $g(s_h, u_1)$

    4. compute $V(s_h, j) = \min(V(s_h, j)$ and $g(s_h, u_1, j) + V1)$

  endfor

Figure 3 shows the procedure in the case L = P.

stage j	0	1	2	3	4	5	6	7
step 0.	s_0, V_0	s_1, V_1	s_2, V_2	s_3, V_3	s_4, V_4	s_5, V_5	s_6, V_6	s_7, V_7
	$S1=s_0$	$S1=s_1$	$S1=s_2$	$S1=s_3$	$S1=s_4$	$S1=s_5$	$S1=s_6$	$S1=s_7$
	$V1=V_0$	$V1=V_1$	$V1=V_2$	$V1=V_3$	$V1=V_4$	$V1=V_5$	$V1=V_6$	$V1=V_7$
step 1.								
shift	$S1=s_1$	$S1=s_2$	$S1=s_3$	$S1=s_4$	$S1=s_5$	$S1=s_6$	$S1=s_7$	$S1=s_0$
	$V1=V_1$	$V1=V_2$	$V1=V_3$	$V1=V_4$	$V1=V_5$	$V1=V_6$	$V1=V_7$,	$V1=V_0$
.	.	.	.	.	.	.	.	.
.	.	.	.	.	.	.	.	.
step L.	$S1=s_7$	$S1=s_0$	$S1=s_1$	$S1=s_2$	$S1=s_3$	$S1=s_4$	$S1=s_5$	$S1=s_6$
	$V1=V_7$	$V1=V_0$	$V1=V_1$	$V1=V_2$	$V1=V_3$	$V1=V_4$	$V1=V_5$	$V1=V_6$

Fig. 3 Procedure P2

In order to complete the computation of the $V(s_h, j)$ for the state s_h at each stage, the processor h must repeat the execution of steps 1 to 4 exactly L times: in this way the processor h receives all the values s_k, $k \neq h$. The time required to execute the steps 1 to 4 is given by

$$t_r + t_b$$

where;

t_r is the time to perform an one step shift: it can be considered a constant;

t_b is the time to execute steps 2 to 4; the hypotesis made on the state transition function f warrants us that this time is constant, as it refers to the computation concerning one and only one value of the decision variable.

Remark that if the function f should not be invertible with respect to u then we could not know a priori the time needed to execute the module, as steps 2 to 4 should be repeated as many times as the number of different decision values corresponding to the pair s_h, s_k.

The total time for procedure P2 is given by:

$$T = N * L * (t_r + t_b)$$

In the case of limited parallelism ($P < L$), each processor is associated to L/P states; after an one step shift the processor repeats steps 2 to 4 for all the L/P states associated to it. The total time will be:

$$T = N * L * L/P * (t_r + t_b)$$

We show now that if we augment the amount of hardware the parallel time decreases by a factor of $0(L/\log L)$. We can implement the same procedure on the orthogonal threes architecture with $L * L$ processors. Here row h is devoted to computations concerning the state s_h, while column k is devoted to computations concerning the state s_k. At each stage the following steps are performed:

Step 1. The values s_k, $V(s_k, j+1)$ are sent to the processors of column k along the column trees starting from the root

Step 2. Each processor h, k computes $u_l = f^{-1}(s_h, s_k, j)$
$$V(s_h, j) = g(s_h, u_l, j) + V(s_k, j+1)$$

Step 3. The minimum over all the values $V(s_h, j)$ is computed along the row tree h; this minimum value is then sent from the root to each processor of the row h.

Step 4. Each processor k, k sends its associated values of s_k and $V(s_k, j+1)$ to the root of the k-th column tree.

Step 1, 3, 4 of the procedure require $0(\log L)$ time, while step 2 requires $0(1)$ time. The total parallel time will be $0(N * \log L)$, that is much better than the time of the one dimensional mesh.

4. Further considerations on procedure P2

In the last section we have seen two parallel procedures for dynamic programming problems:

— the first procedure can solve any dynamic programming problem on a two dimensional mesh with $L * L$ processors in time $0(N * M * L)$; it does not reach the optimal speed-up ratio because of the overhead for the communication

— the second procedure can solve dynamic programming problems, whose state transition function f is invertible with respect to u, in time $0(N * \log L)$ in an orthogonal trees architecture with $L * L$ proceessors and in time $0(N*L)$ in an one dimensional mesh with L processors.

Many dynamic programming problems satisfy the hypothesis on the invertibility of f, for instance all the problems whose f is linear with respect to u: however procedure P2 is not always the more convenient in terms of the total time. If $M \ll L$ in fact the procedure

P1 on the mesh is better than the procedure P2 on the array; moreover P1 is not significantly worse than P2 on the orthogonal trees, mostly if we take into account that we use much less hardware (see table 1).

An example of this situation is given by the binary knapsack problem; in this case the cardinality M of the decision space is two, that is $M \ll L$; moreover the communication procedure is activated only once at each stage, for the value 1 of the decision variable; by comparing the total time of the two procedures, P1 results to be always better than P2, as $M < \sqrt{L}$.

An example of problem for which P2 works better than P1 is given by the following resource allocation problem: there are M units of resource to be allocated to N projects. A net profit $C(u_1, j)$ is associated to the allocation of u_1 units of resource $(0 < u_1 < M)$ to project j.

The problem is to allocate the resource to the N projects in such a way as to maximize the total net profit.

At each stage j we have to solve the following maximization problem:

$$V(s_h, j) = \max (V(s_h - u_1, j\text{-}1) + C(u_1, j)$$

where s, the state variable, represents the number of units of resource already allocated to projects 1, 2 . . . j-1; the state transition function:

$$s_k = s_h - u$$

is invertible with respect to u.

At each stage the number of feasible values of the decision variable varies from 0 to M, depending on the state values; on the other hand the state variable takes the values from 0 to M; therefore $L = M$ and the procedure P2 results to be more convenient.

In table 1 the evaluation of the two procedures is given in terms of computation time, number of processors and the product (time * number of processors).

The last column of table 1 contains the evaluation of P1 and P2, implemented in VLSI technology, in terms of the AT^2 measure of complexity: the comparison of the values brings to the same considerations as before.

procedure	architecture	time	n. of proc.	time∗n. of. proc.	AT^2
P1	mesh	$N*M*\sqrt{L}$	L	$N*M*L*\sqrt{L}$	$N*M*L^{3/2}$
	orthogonal trees	$N*L$	$L*M$	$N*L^2*M$	N^2*L^3*M* logL∗logM
P2	array	$N*L$	L	$N*L^2$	N^2*L^3
	orthogonal trees	$N*logL$	$L*L$	$N*L^2*$ log L	$N^2*L^2*logL^3$

Table 1

References

[1] Dano : *"Non linear and dynamic programming,"* Addison Wesley, 1975

[2] Hadley: *"Non linear and dynamic programming,"* Springer Werlag, 1964.

[3] R. E. Larson, J. L. Casti: *"Principles of dynamic programming,"* M. Dekker, 1978.

[4] H. T. Kung: *"The structure of parallel algorithms,"* Advances in Computer, Marshall C. Yovits, 1980.

[5] B. Lint et al. : *"Communication issues in the design and analysis of parallel algorithm".* IEEE Trans. on Soft. Eng., Vol. 7, n. 2, March 1981.

[6] J. L. Casti, Richardson, R. E. Larson: *"Dynamic programming and parallel computers,"* J.O.T.A., Vol 12, n. 4, 1973.

[7] Al–Dabass: *"Two methods for the solution of the dynamic programming algorithm on a multiprocessor cluster",* Optimal Control Application and Methods, vol. 1, 1980.

[8] L. J. Guibas et al. : *"Direct VLSI implementation of combinatorial algorithms",* VLSI Conference, Caltech 1979.

[9] D. P. Bertsekas: *"Distributed dynamic programming",* IEEE Trans. on Aut. Control, Vol 27, pp. 610-616, 1982.

[10] F. T. Leighton : *"New lower bound techniques for VLSI"*, Proc. 22-th IEEE Symp. on Found. of Comp. Sci., Nashville, Tenn., Oct. 1981.

[11] D. Nassimi, S. Sahni: *"Data broadcasting in SIMD computer"*, IEEE Trans. on Comp., vol. C-30, Feb. 1981.

[12] C. D. Thompson, H. T. Kung: *"Sorting on a mesh connected parallel computer"*, Comm. ACM, 20-4, 1977.

[13] U. Montanari: *"On optimal detection of curves in noisy pictures"* Comm. ACM, vol. 14, May 1974.

[14] P. Bertolazzi, M. Pirozzi: *"Two parallel algorithms for dynamic programming"*, II World Conf. On Math. at the Service of Man, Las Palmas, June 1982.

Annals of Discrete Mathematics 25 (1985) 65-90
© Elsevier Science Publishers B.V. (North-Holland)

SIMULATIONS AMONG CLASSES OF RANDOM ACCESS MACHINES AND EQUIVALENCE AMONG NUMBERS SUCCINCTLY REPRESENTED[1]

A. BERTONI - G. MAURI - N. SABADINI

Istituto di Cibernetica - Università di Milano

1. Introduction

The first part of this paper is concerned with the analysis of the computational complexity of enumeration problems and with the comparison among the complexity classes with respect to Counting Turing Machines, as defined by Valiant [11, 12], and the complexity classes with respect to Random Access Machines.

Enumeration problems constitute a major part of combinatorial mathematics. The input of an enumeration problem is the description of a (finite) set; the required output is the cardinality of such a set.

In order to deal with the question of the computational complexity of enumeration problems, Valiant [11, 12] introduced the notion of Counting Turing Machine, a generalization of the Non Deterministic Turing Machine, as a formal model for describing enumeration problems, and studied the class of enumeration problems complete in polynomial time with respect to the given model. An equivalent approach has been followed by Simon [7, 8], who defined the Threshold Machines.

On the other hand, combinatorial mathematics expresses the solution of enumeration problems by means of solving formulas, generally based on the usual arithmetic operations [4]. Since these formulas can

---

[1] *A preliminary version of the first part of this paper has been published in the Proceedings of the ACM Symp. on Theory of Computation, Milwaukee, 1981*

be formally represented as programs for a Random Access Machine (RAM) with arithmetical primitives, the natural complexity measure is the arithmetic complexity.

Starting from this consideration, here we consider the classes of enumeration problems with respect to their arithmetical complexity, and compare the complexity classes so obtained with the classes on Turing Machines.

First, the class # P–SPACE of enumeration problems solvable in polynomial space by a Counting Turing Machine is defined and a problem complete in this class is shown, i.e. the enumeration problem associated with the Satisfiability Problem for Quantified Boolean Formulas.

Then, we prove a strong characterization theorem, stating that every problem in # P–SPACE can be solved in polynomial time by a RAM with the operations of sum, product, integer subtraction and integer division. The proof uses some results on the efficient evaluation of polynomials that are given in Section 5.

In the second part of the paper, we present two interesting consequences of the main result. The first one refers to the problem of the simulability among classes of Random Access Machines with different sets of operations (including Vector Machines [5] as RAM's with boolean operations). This problem has been studied by Hartmanis and Simon [3], Simon [7, 9], Pratt and Stockmeyer [5] and Schönage [6]. In particular, the latter paper gives an almost complete comparison among the different classes of RAM's with respect to their computational power. By using our result, we are able to give an answer to a well known open problem: RAM's with the set of operations $\{+, \div, *, \div\}$ can polynomially simulate Vector Machines.

Second, we prove that, given a polynomial time arithmetical RAM computing a function f: $\Sigma^* \longrightarrow N$, there is a polynomial time deterministic Turing Machine which, on input x, prints out a straigth line program with arithmetic operations which represents in a succinct form the number f(x). This is a rather surprising result, since we have proved that polynomial time on RAM's is equivalent to polynomial space on TM's; hence, we are induced to suspect that the usual relation of equality on the set of natural numbers represented as arithmetic straigth line programs is difficult to compute; in fact, we prove that this problem is P–SPACE complete. A similar problem was studies by Schönage [6] for the particular case of straigth line programs using only the operations + and *. In this case, the equivalence problem turns out to be co–NP. Our result, compared with the one of Schönage, gives a strong evidence to Simon's conjecture [9] on the power of integer division with respect to the other arithmetic operations.

2. Complexity classes of enumeration problems

Usually, non deterministic and alternating TM's are used only to solve decision problems. However, Valiant [11, 12] and Simon [7, 8] carried out a generalization by considering enumeration problems. Roughly speaking, while decision problems consist in establishing if a structure with a particular property exists in a set of given structures, an enumeration problem consists in counting the number of structures with a given property.

A formal setting for the concept of enumeration — that is fundamental in combinatorial theory [4] — can be established as follows.

Definition 2.1. An *enumeration system* is a pair $\langle I, \mathcal{J} \rangle$ where I is an index set and $\mathcal{J}$ is a system of finite sets indexed by I: $\mathcal{J} = \{ S_j | j \in I \}$.

Definition 2.2. The *enumeration function* associated with an enumeration system $\langle I, \mathcal{J} \rangle$ is the function f: $I \longrightarrow N$ defined by $f(j) = \#S_j$. (Here, N denotes the set of natural numbers and $\#S$ the cardinality of the set S).

Some different computational models for the description of enumeration systems (up to encoding questions) have been given: for example, Counting Turing Machines in Valiant [12] and Threshold Machines in Simon [7]. The description by means of Counting Turing Machines is based on the fact that every non deterministic Turing machine has in general a number of different accepting paths for every accepted input, so that we can associate with it an enumeration system as follows:

Definition 2.3. An enumeration system $\langle I, \mathcal{J} \rangle$ is *generated by a non deterministic Turing machine M* with input alphabet Σ iff $I = \Sigma^*$ and for every $w \in \Sigma^*$, S_w is the set of (different) accepting computations of M on input w.

Definition 2.4. A *Counting Turing Machine (CTM)* is a non deterministic Turing machine which (magically) prints the number $\#S_w$ for every input w; so, it computes the enumeration function of the generated system $\langle I, \mathcal{J} \rangle$.

In [11], Valiant defines the notion of time complexity of a CTM and the class of #P—complete problems. We can extend these definitions to cover also space complexity as follows.

Definition 2.5. A Counting Turing Machine M has:

— *time complexity* T(n) iff for every input w of length n the *longest* computation accepting w requires at most T(n) steps;

— *space complexity* S(n) iff for every input w of length n every computation:

   1) requires at most S(n) cells; and

   2) halts at most after $2^{n \cdot S(n)}$ steps.

Condition (2) identifies a particular subclass of the class of CTM's with space bound S(n). Since $2^{n \cdot S(n)}$ is an upper bound for the number of different instantaneous descriptions of a S(n)—space bounded Turing machine, $2^{n \cdot S(n)}$ steps are sufficient to capture all the computations without repetitions of instantaneous descriptions. Furthermore, the given bound allows to exclude machines for which the total number of accepting computations can exceed $2^{2^{S(n)}}$, and in particular be infinite. Such a limitation, needed in order to keep the power of Counting Turing Machines within reasonable bounds, is analogous to the limitations imposed by Pratt and Stockmeyer [10] in order to forbid computations of Vector Machines in which the vectors grow too large.

Definition 2.6. #P and #P—SPACE are the classes of enumeration problems that can be solved by a CTM working in polynomial time or, respectively, space.

While, as far as complexity classes of decision problems are concerned, it is not known if P−SPACE = NP, for enumeration problems the fact that #P−SPACE $\supsetneq$ #P follows from the obvious combinatorial remark that in the class #P−SPACE are contained some problems with 2^{2^n} solutions, being n the size of the input, while for every problem in #P there exists a polynomial P(n) such that the number of solutions is less than $2^{P(n)}$.

3. A #P−SPACE complete problem

The notions of polynomial reducibility and of completeness cannot be directly extended to the classes #P and #P−SPACE. In fact, we have to require that the polynomial reduction preserves not only the existence or the absence of accepting paths, but preserves their exact number, i.e. that it be *parsimonious* [11].

It is in general easy to prove that the enumeration problem associated with a NP−complete problem is #P−complete. However, in [12] Valiant exhibits a #P−complete problem whose associate decision problem is in P.

Theorem − Let (A_{ij}) be a 0,1 square matrix of dimension n, and let the permanent $Perm(A_{ij})$ be defined by:

$$Perm(A_{ij}) = \sum_p \prod_i A_{p(i),i}$$

where p denotes a permutation of the n-tuple (1,..., n). Then the problem "Calculus of the permanent of A" is a #P−complete problem.

Computing the permanent of A corresponds to counting the number of perfect matchings of the associated graph: the relevant fact is that we can decide on the existence of at least one perfect matching in polynomial time.

In order to exhibit a problem complete in #P−SPACE, we can consider the enumeration problem associated with a decision problem which has been shown to be P−SPACE−complete in [10], i.e. the satisfiability problem for quantified Boolean formulas.

Definition 3.1. A *quantified booleans formula* (QBF) is a first order boolean sentence of the form:

$$\psi = Q_1 \, x_1 \, Q_2 \, x_2 \, \cdots \, Q_m \, x_m \, \varphi \, (x_1 \cdots x_m)$$

where $Q_1 \in \{\forall, \exists\}$ and φ is a quantifier-free boolean formula. The variable x_j is said to be *universal* iff $Q_j = \forall$, *existential* iff $Q_j = \exists$.

Definition 3.2. Let $Q = Q_1 \cdots Q_m$ be a sequence of quantifiers $Q_j \in \{\forall, \exists\}$. The set $\mathscr{D}(Q)$ of the *assignment trees generated by Q* is recursively defined by:

$$\mathscr{D}(\forall) = \{\{0,1\}\}$$
$$\mathscr{D}(\exists) = \{\{0\},\{1\}\}$$
$$\mathscr{D}(\forall Q) = \{0.T_1 \cup 1.T_2 \mid T_1, T_2 \in \mathscr{D}(Q)\}$$
$$\mathscr{D}(\exists Q) = \{0.T \mid T \in \mathscr{D}(Q)\} \cup \{1.T \mid T \in \mathscr{D}(Q)\}$$

where obviously $a.T = \{at \mid t \in T\}$, for $a \in \{0,1\}$

Intuitively, a tree $T \in \mathscr{D}(Q_1 \cdots Q_m)$ is a set of strings $t = t_1 \cdots t_m$, with $t_j \in \{0,1\}$, that can be interpreted as assignments to the variables x_j.

Definition 3.3. Let $\psi = Q_1 x_1 \cdots Q_m x_m \, \varphi(x_1 \cdots x_m)$ be a QBF; a tree $T \in \mathscr{D}(Q_1 \cdots Q_m)$ is *accepting* with respect to ψ iff for every $t = t_1 \cdots t_m \in T$ we have $\psi(t_1 \cdots t_m) = 1$.

Now, we define the following problem:

SATISFIABILITY OF QUANTIFIED BOOLEAN FORMULAS (#QBF)

INSTANCE: A well formed QBF $\psi = Q_1 x_1 \cdots Q_n x_n \, \varphi \, (x_1 \cdots x_n)$ where φ is a boolean formula involving the variables $x_1, \cdots x_n$, and each Q_j is either $\forall$ or $\exists$.

QUESTION: Find the number of different assignment trees accepting with respect to ψ.

In [10] it is shown that the corresponding decision problem is complete in P−SPACE. Since the reduction used in [10] is not parsimonious, the proof cannot be immediately extended to the counting problem in order to show its completeness. However, the proof can be rewritten by using a parsimonious reduction, hence proving the following:

Theorem 3.1. # QBF is complete in #P—SPACE.

Proof —Let M be a CTM which accepts some language $L \in \{0,1\}^*$ in space $\leqslant S(|x|)$ ($|x|$ being the lenght of the string x), for some polynomial S. Let us now consider a Turing Machine M' which simulates M until M enters a final configuration, then repeats this configuration. So, the number of accepting computations of M' of length (number of steps) $2^{|x|S(|x|)}$ is just the number of accepting computations of M.

Furthermore, in the computation of M' on input x, no instantaneous description can exceed the length $1+S(|x|)$. Now, we choose an encoding of states and symbols of M' into words in $\{0,1\}^*$ so that every instantaneous description of M' during its computation on input x will be encoded as a word $y \in \{0,1\}^*$ such that $|y| \leqslant q(|x|)$ for some polynomial q not depending on x. In particular, 1 will denote the instantaneous description code in which only the symbol 1 appears.

As in [10], one can construct a boolean formula $A(u,v)$, where u, v are sequences of boolean variables, of length $n = q(|x|)$, such that:

$A(u,v) \Longleftrightarrow u$ and v are the encoding of i.d. 's of M' and v follows from u in one step.

Moreover, for fixed M, the length of A is bounded by a polynomial in n.

We are interested in building a quantified boolean formula $\mathscr{C}_k(u,v)$ which holds iff v is an i.d. of M' obtained from u in 2^k steps, and such that the number of assignment trees accepting with respect to $\mathscr{C}_k(u,v)$ is the number of different derivations, in 2^k steps, of v from u. Let:

$$A_{j+1}(x_{j+1}, y_{j+1}, ..., z_k, x_k, y_k, ... z_0, x_0, y_0)$$

be defined as follows:

$$A_1(x_1, y_1, z_0, x_0, y_0) = (C_1 \wedge A(x_0, y_0)) \vee \sim C_1$$

$$A_{j+1} = (C_{j+1} \wedge A_j) \vee (\sim C_{j+1} \wedge B_j)$$

where:

$$C_{j+1} = (x_{j+1} = x_j \wedge z_j = y_j) \vee (x_j = z_j \wedge y_j = y_{j+1})$$

$$B_j = (z_{j-1} = 1) \wedge ... \wedge (z_1 = 1) \wedge (z_0 = 1)$$

Now, by using the convention $Q_i = \exists z_i \; \forall x_i \; \forall y_i$, we define:

$$\mathscr{C}_{j+1} (x_{j+1}, y_{j+1}) = Q_j \, Q_{j-1} \cdots Q_0 \, A_{j+1}$$

If $j = |x|S(|x|)$, then u is the initial configuration and, being fin(v) the formula defined by:

$$\text{fin } (v) \Longleftrightarrow v \text{ is a final configuration in } M$$

we consider the formula:

$$\psi = \exists z \, \exists v (\mathscr{C}_j (z, y) \wedge z = u \wedge \text{fin } (y))$$

The length of this formula is bounded by a polynomial in $|x|$, and ψ can be easily built up in polynomial time with respect to $|x|$.

Now, we will show that the number of assignment trees accepting with respect to ψ is the number of accepting computations of M on input x.

The function #, which associates to every QBF the number of accepting assignment trees, verifies the following recurrence relation for $k_i \in \{ \forall, \exists \}$:

$$\# \, k_1 \, x_1 \, ... \, k_n \, x_n \, \varphi(x_1 \, ... \, x_n) =$$

$$\text{If } k_1 = \exists \text{ then} \sum_{x_1} \# \, k_2 \, x_2 \, ... \, k_n \, x_n \, |\varphi(x_1, \, ..., \, x_n)$$

$$\text{else } \prod_{x_1} \, \# k_2 \, x_2 \, ... \, k_n \, x_n \, \varphi(x_1, \, ..., \, x_n)$$

So, by induction techniques, we are able to prove:

a) $\# \, Q_{j-1} \, Q_{j-2} \cdots Q_0 B_j = 1$

In fact, $\# Q_0 B_1 = 1$ can be easily verified. Moreover, we have:

$$\#Q_{j-1} \cdots Q_0 B_j = \sum_{z_{j-1}} \prod_{x_{j-1}} \prod_{y_{j-1}} \#Q_{j-2}(z_{j-1} = 1 \wedge B_{j-1})$$

$$= \prod_{x_{j-1}} \prod_{y_{j-1}} Q_{j-2}(B_{j-1}) = \prod_{x_{j-1}} \prod_{y_{j-1}} 1 = 1, \text{ by induction hypothesis.}$$

b) $\# \mathcal{C}_{j+1}(x_{j+1}, y_{j+1})$ is the number of derivations by M' in 2^{j+1} steps of the i.d. y_{j+1} from the i.d. x_{j+1}.

In fact, the case $j = 0$ can be immediately verified. Moreover, we have:

$$\# \mathcal{C}_{j+1} = \sum_{z_j} \prod_{y_j} \prod_{y_j} \#Q_{j-1} \cdots Q_0 (C_{j+1} \wedge A_j) \vee (\sim C_{j+1} \wedge B_j) =$$

$$= \sum_{z_j} \prod_{x_j} \prod_{y_j} Q_{j-1} \cdots Q_0 (\text{If } C_{j+1} \text{ then } A_j \text{ else } B_j) =$$

$$= \sum_{z_j} \prod_{x_j} \prod_{y_j} \text{If } C_{j+1} \text{ then } \#Q_{j-1} \cdots Q_0 A_j \text{ else } \#Q_{j-1} \cdots Q_0 B_j =$$

$$= \sum_{z_j} \prod_{x_j} \prod_{y_j} \text{If } C_{j+1} \text{ then } \# \mathcal{C}_j \text{ else } 1 =$$

$$= \sum_{z_j} \# \mathcal{C}_j (x_{j+1}, z_j) \cdot \# \mathcal{C}_j (z_j, y_{j+1})$$

By induction hypothesis, $\# \mathcal{C}_j(x_{j+1}, z_j)$ (resp. $\# \mathcal{C}_j(z_j, y_{j+1})$) is the number of derivations by M' in 2^j steps of z_j from x_{j+1} (resp. of y_{j+1} from z_j). Then, $\# \mathcal{C}(x_{j+1}, y_{j+1})$ is the number of derivations by M' in 2^{j+1} steps of y_{j+1} from x_{j+1}. Hence:

$$\# \psi = \# \exists z \exists y (\mathcal{C}(z, y) \wedge z = u \wedge \text{fin } (y))$$

is the number of accepting computations of M' on input x in 2^j steps ($j = |x|.S(|x|)$). So, we can conclude that $\#\psi$ is the number of accepting computations of M on input x.

4. Random access machines and vector machines

The complexity of decision problems has been studied also with respect to models of computation different from Turing Machines, in particular Vector Machines [5] and Random Access Machines [6], which embed implicit parallel features. Before discussing the power of RAM's in solving enumeration problems, we will now give some definitions and previous results.

Definition 4.1 A basic *Random Access Machine (RAM)* consists of:

a) a *memory* consisting of an enumerable set of registers, with addresses 0, 1, ..., n, ..., each of which can contain an integer of arbitrary size; the current contents of the register n are denoted by $\langle n \rangle$; register 0 is used as accumulator.

b) a *program* P, consisting of a finite sequence of labeled instructions. A table of the admitted instructions, together with their semantics follows; the notation $\langle n \rangle := \langle k \rangle$ is used to mean that the contents of k are copied in n; $\langle\langle n \rangle\rangle$ will indicate the contents of the register whose address is the contents of n: this procedure will be referred to as "indirect addressing".

Instructions		Semantics
a-load	n	$\langle 0 \rangle := n$
load	n	$\langle 0 \rangle := \langle n \rangle$
i-load	n	$\langle 0 \rangle := \langle\langle n \rangle\rangle$
store	n	$\langle n \rangle := \langle 0 \rangle$
i-store	n	$\langle\langle n \rangle\rangle := \langle 0 \rangle$
suc		$\langle 0 \rangle := \langle 0 \rangle + 1$
goto	λ	The control is transferred to λ
jzero	λ	If $\langle 0 \rangle = 0$, the control is transferred to λ
halt		The machine stops

Definition 4.2. A *computation* of a RAM is as follows:

a) we start with an input $\langle x_1 \ldots x_h, h \rangle$, where $x_1 \ldots x_h \in \{0,1\}^*$ gives the *"structure"* of the input, and $h = |x_1 \ldots x_h|$ gives its *size*. The input is stored as follows:

$\langle 1 \rangle = h$

$\langle j \rangle = x_{j-1}$ for $1 \leqslant j \leqslant h+1$

$\langle j \rangle = 0$ for $j > h+1$

b) the program is executed sequentially if no jump *(goto)* or branching *(jzero)* instructions are encountered.

c) a special *output register* z is distinguished. Its content after the
machine stopped is the result of the computation.

Definition 4.3. A RAM M has time complexity $T(n)$ if the longest
computation originated by any input of lenght n requires to execute
$T(n)$ instructions.

More powerful RAM's can be obtained by adding extra instructions
from the following list:

add	n ; +	$\langle 0 \rangle := \langle 0 \rangle + \langle n \rangle$;
sub	n ; ∸	$\langle 0 \rangle := \max\{0, \langle 0 \rangle - \langle n \rangle\}$;
mult	n ; *	$\langle 0 \rangle := \langle 0 \rangle * \langle n \rangle$;
div	n ; ÷	$\langle 0 \rangle := \langle 0 \rangle / \langle n \rangle$; *halt,* if $\langle n \rangle = 0$;
shift	n ; ←	$\langle 0 \rangle := \langle 0 \rangle / 2^{\langle n \rangle}$;
and	n ; ∧	$\langle 0 \rangle := \langle 0 \rangle \wedge \langle n \rangle$, where both operands are considered as binary strings.

In the following, the class of RAM's with the set $\{o_1, ..., o_n\}$ of additional instructions will be denoted by RAM $(o_1, ..., o_n)$.

Vector Machines (VM's) [5] are a variant of RAM's in which the registers can contain arbitrarily long sequences of bits (elements of $\{0,1\}$)
rather than arbitrarily large numbers. The following operations can be
executed on such "bit vectors":

A ← K   the constant bit-vector K is loaded into register A
A ← $\overline{B}$ "bitwise parallel" boolean complement
A ← B∧C "bitwise parallel" boolean conjunction
A ← B↓C left shift of B by the distance given by C
A ← B↑C right shift of B by the distance given by C
A = 0   predicate for testing whether A is 0 everywhere.

It is evident that VM's make use of unbounded parallelism, since
one computation step (one operation) consists of the execution of an
arbitrarily large number of bitwise operations. An analogous remark

holds for RAM's, since it is assumed that an operation has unit cost, independently of the value of operands. These remarks account for the power of RAM's and VM's. In particular, in [5] it is proved that:

$$V_k\text{--PTIME} = \text{P--SPACE}$$

where V_k is the subclass of vector machines for which at the k-th step in a computation the maximum length of vectors is bounded by 2^{k+n}, where n is the size of the initial vector.

We can now compare classes of RAM's with different sets of operations by using the notion of polynomial reducibility.

Definition 4.4. A class $\mathscr{R}_1$ of RAM's is *polynomially reducibile* to the class $\mathscr{R}_2$, denoted by $\mathscr{R}_1 \xrightarrow{P} \mathscr{R}_2$, iff for every machine $M_1 \in \mathscr{R}_1$ there exists a $M_2 \in \mathscr{R}_2$ such that M_2 polynomially simulates M_1.

Some results on the polynomial reducibility among different classes of RAM'S, given in [5, 6], are collected in the following diagram:

$$V_k \xleftarrow{\;\;P\;\;} \text{RAM}(+, \div, *, \wedge) \xleftarrow{\;\;P\;\;} \text{RAM}(+, *, \wedge)$$

$$\uparrow \qquad\qquad\qquad \uparrow \qquad\qquad\qquad \uparrow$$

$$\text{RAM}(+, \div, *, \div) \xrightarrow{\;\;P\;\;} \text{RAM}(+, \div, *, \downarrow) \xleftarrow{\;\;P\;\;} \text{RAM}(+, *, \downarrow)$$

In the next sections we will show that also the classes of machines on the first row can be polynomially reduced to those on the second row, so completing the diagram.

5. Efficient evaluation of polynomials represented as straigth line programs

Arithmetical RAM's are a very natural tool for solving enumeration problems represented by CTM's, since the "auxiliary device" which computes the counting function can be implemented by a program for a RAM. In fact, combinatorial mathematics often expresses the solution of an enumeration problem by means of a "solving formula" based on the usual arithmetic operations. Two examples of such solving formulas are as follows:

Example 1 – # {Permutations of $\{1, 2, ..., n\}$} $= \prod_{1}^{n} k = n!$

Example 2 – # {Cycle decompositions of a graph with characteristic matrix A} $=$ perm$(A) = \Sigma A_{1k_1} A_{2k_2} \cdots A_{nk_n}$, where the sum is extended over all the permutations of the set $\{1, 2, ..., n\}$.

Hence, it is interesting to classify enumeration problems with respect to their complexity in time on RAM's and to compare the classes so obtained with the complexity classes on Turing Machines. The main theorem given in the following section is based on the fact that #QBF can be solved in polynomial time on a RAM with the set of arithmetic operations $\{+, \div, *, \div\}$.

Since the proof of this fact will use high level computations on polynomials (generating functions), as often happens for enumeration problems, in this section we will give some preliminary results on the efficient evaluation of polynomials. To be more formal, we will consider the following two structures:

a) The structure

$$\mathcal{N} = \ <N, +, \div, *, \div>$$

of integer non negative numbers, with the usual arithmetic operations of sum, product, non negative difference (i.e. $n \div m = $ Max $(0, n - m)$) and integer division. Furthermore, we will use the derived operation

$$x \bmod y = x \div y * (x \div y)$$

b) The structure

$$\mathcal{P} = \ <P, +, ., \otimes, |, \boxed{h}>$$

where P is the set of polynomials in a single variable with integer non negative coefficients; here, a polynomial of degree n is intended to be a function $f: [n+1] \to N$, where $[n+1] = \{0, 1, ..., n\}$.

The number $f(k)$, $0 \leqslant k \leqslant n$, is called the k-th coefficient of f, and denoted by f_k.

Let now f and g be two polynomials of the same degree n. The operations on the structure are defined as follows:

$$(f+g)_k = f_k + g_k \qquad (0 \leqslant k \leqslant n)$$

$$(f.g)_k = \sum_{j=o}^{k} f_j {}^* g_{k-j} \qquad (0 \leqslant k \leqslant 2n)$$

$$(f \otimes g)_k = f_k {}^* g_k \qquad (0 \leqslant k \leqslant n)$$

$$(|f)_k = f_{k+1} \qquad (0 \leqslant k \leqslant n-1)$$

$$(\boxed{h} \, f)_k = f_{hk} \qquad (0 \leqslant k \leqslant \lfloor n/h \rfloor)$$

These operations can be efficiently evaluated in terms of arithmetic operations $+, \div, {}^*, \div$, as stated by the following theorem.

Theorem 5.1. Given the values, x, z, f(x), g(x), z^{n+1}, with $x > f_o$ and $z > f(x) {}^* g(x^{n+1}) {}^* (n+2)$, the values $(f+g)(x)$, $(f.g)(x)$, $(f \otimes g)(x)$, $(|f)(x)$ and $(\boxed{h} \, f)(x)$ can be computed with $0(1)$ arithmetic operations.

Proof –

a) Operations $+$ and $\cdot$

   Immediate, from the definition:

   $(f+g)(x) = f(x) + g(x)$ and $(f.g)(x) = f(x) {}^* g(x)$

b) Operation $|$

   $(|f)(x) = f(x) \div x$

   In fact, since $x > f_o$ implies $f_o \div x = 0$, we have:

   $$f(x) \div x = (\sum_{k=o}^{n} f_k x^k) \div x = \sum_{k=1}^{n} f_k x^{k-1} + f_o \div x = (|f)(x)$$

c) Operation $\boxed{h}$

   $(\boxed{h} \, f)(x) = (f(z) \bmod z^h - x) \bmod z$

We give the idea in the case $h = 1$. For $h > 1$ the proof is similar.

$$(f(z) \bmod (z\text{-}x)) \bmod z = f(z) \bmod (z\text{-}x) = \sum_{k=o}^{n} f_k z^k \bmod (z\text{-}x) =$$

$$= (\sum_{k=o}^{n} f_k (z^k\text{-}x^k) + \sum_{k=o}^{n} f_k x^k) \bmod (z\text{-}k) = \sum_{k=o}^{n} f_k x^k \bmod (z\text{-}k) =$$

$$= f(x) \bmod (z\text{-}x)$$

Finally, from the hypothesis $f(x) + x < 2f(x) < z$, it follows that:
$f(x) \bmod (z\text{-}x) = f(x) = (\boxed{1}\, f)\, (x)$.

d) Operation $\otimes$

$(f \otimes g)\, (x) = (f(z)g(z^{n+1}) \bmod (z^{n+2}\text{-}x) \bmod z$

In fact, we have:

$$f(z)\, g\, (z^{n+1}) = \sum_{j=o}^{n} \sum_{k=o}^{n} f_j g_k z^{j+k(n+1)}$$

so that $f(z)g(z^{n+1})$ can be considered to be the value in z of a

new polynomial $q = \sum_{i=o}^{n^2+2n} q_i z^i$ of degree $n^2 + 2n$, where $q_i = f_j g_k$,

with: $j = i \bmod (n+1)$ and $k = i \div (n+1)$. Thus, from c), we have:

$$(q(z) \bmod (z^{n+2}\text{-}x)) \bmod z = \sum_{k} q_{(n+2)k} x^k$$

But $q_{(n+2)k} = q_{(n+1)k+k} = f_k g_k = (f \otimes g)_k$, from which the thesis follows.

Now, we can think of the polynomial in P as generated by the constant 1 (i.e. $f_o = 1$ and $f_i = 0$ for $i \neq 0$), the elementary polynomial $\lambda\, x(x)$ (i.e. $f_1 = 1$ and $f_i = 0$ for $i \neq 1$) and a finite sequence of the operations $+,\, .,\, \otimes,\, \boxed{2},\, |.$

In the same way, numbers can be seen as generated by the constants 0 and 1 by application of a finite sequence of arithmetical operations. This corresponds to giving a symbolic representation of polynomials and numbers, as follows.

Definition 5.1. A *straight-line program* (slp) of length n on the structure $\mathscr{P}$ (respectively, on the structure $\mathscr{N}$) is a sequence of n instructions with labels, in order, (1), ..., (n) such that:

a) The first instruction is

$$(1) \quad \text{pl} \longleftarrow \quad \lambda \, x(x)$$
$$(x1 \longleftarrow \quad 0/1)$$

b) the kth instruction has one of the following (alternative) forms:

$$(k) \quad \text{pk} \longleftarrow \text{pj+ps / pj.ps / pj} \otimes \text{ps /|ps/} \; \boxed{2} \; \text{ps / 1}$$
$$(xk \longleftarrow \text{xs+xj / xs} \div \text{xj / xs*xj / xs} \div \text{xj / xs})$$

with $s, j < k$.

Definition 5.2. Let Π be a slp on $\mathscr{P}$ of length n; it *generates* the polynomial $p(\Pi) := pn$.

A bound for the degree of $p(\Pi)$ with respect to the length n of the slp Π is given by the following:

Fact 5.2. Let $p(\Pi)$ be the polynomial generated by a slp Π of length n; then:

a) degree $(p(\Pi)) \leqslant 2^n - 1$;

b) Max $p_k \leqslant 2^{3^n}$

where p_k is the coefficient of x^k in $p(\Pi)$.

Definition 5.3. Let ϱ be a slp on $\mathscr{N}$ of length n; it has the *value* $v(\varrho) := xn$.

Fact 5.3. Let $v(\varrho)$ be the value of a slp ϱ on $\mathscr{N}$ of length n. Then:

$$v(\varrho) \leqslant 2^{2^n}$$

We can now state the main results of this section.

Theorem 5.3. Let p be generated by a slp Π of length n; then, given $x \geqslant 2$, the value $p(x)$ can be obtained with $0(n^2)$ arithmetic operations.

Proof — The algorithm that computes p(x) is as follows:

Input: a slp Π ; an integer n (the length of Π); an integer $x \geqslant 2$.

Step 1: we compute the sequence $(x_n, x_{n-1}, ..., x_1)$ defined by the recursion:

$$x_n = x$$
$$x_k = (2x_{k+1})^{16^n} \qquad \text{for } 1 \leqslant k \leqslant n$$

Time: $0\,(n^2)$

Step 2: we compute the sequence $(y_{n-1}, ..., y_1)$ by $y_k = 2x_k^{8^n} -1$ for $1 \leqslant k \leqslant n$
Time: $0(n^2)$

Step 3: we compute the sequence $(pl(x_1), p2(x_2), ..., pn(x_n))$ where pj is the polynomial generated by Π restricted to the first j instruction as follows:

a) $pl(x_1) = x_1$
b) for $2 \leqslant k \leqslant n$, do the following computations:

1) if the kth instruction is pk $\longleftarrow$ pj op ps, op $\in \{+,.\}$, then:

$$pj(x_k) = pj(x_j) \bmod (x_j - x_k)$$
$$ps(x_k) = ps(x_s) \bmod (x_s - x_k)$$
$$pk(x_k) = pj(x_k) \text{ op } ps(x_k)$$

The correctness of the computation is guaranteed by Th. 5.1, a).

2) if the kth instruction is pk $\longleftarrow$ pj $\otimes$ ps, then:

$$pj(y_k) = pj(x_j) \bmod (x_j - y_k)$$
$$ps(y_k) = ps(x_s) \bmod (x_s - y_k^{2^n})$$
$$pk(x_k) = (pj\ (y_k) \cdot ps\ (y_k^{2^n}) \bmod (y^{2+1}-x_k)) \bmod y_k$$

The correctness follows from Th. 5.1, d), by observing that

$2^n > \deg(pj)$ and $2^n > \deg(ps)$ and that $y_k > pj(x_k).ps(x_k^{2^n})$.
(2^n+1).

3) if the kth instruction is ph $\longleftarrow$ | pj, then:

$$pj(x_k) = pj(x_j) \bmod (x_j - x_k)$$
$$pk(x_k) = pj(x_k) \div x_k$$
by using Th. 5.1, b).

4) if the kth instruction is pk $\longleftarrow$ $\boxed{h}$ pj, $h \in \{1, 2\}$, then:

$$pj(y_k) = pj(x_j) \bmod (x_j - y_k)$$
$$pk(x_k) = (pj(y_k) \bmod (y_k^h - x_k)) \bmod y_k$$
by using Th. 5.1, c).

Time: $O(n^2)$

Output: an integer $pn(x_n) = p(x)$

We can restate the above theorem in a form where the value $p(x)$ is not explicitly given in one of the usual representations (i.e. decimal or binary), but is represented in symbolic form as a slp on $\mathcal{N}$.

Theorem 5.4. There is a polynomial time deterministic Turing machine which, when receives as input a pair $<\widehat{\Pi}, x>$ where $\widehat{\Pi}$ is a slp on $\mathcal{P}$ and $x \in N$, $x \geqslant 2$, computes a slp ϱ_x on $\mathcal{N}$ whose value is the number $p(x)$.

The proof is as the one of Th. 5.3, where arithmetical operations are not executed, but are written in symbolic form.

6. A characterization theorem

In this section, we will show our main result, i.e. the class of #P−SPACE functions coincides with the class of functions computable in polynomial time on arithmetical RAM's. As a consequence, we will

be able to prove that Vector Machines [5] can be simulated in polynomial time by arithmetical RAM's.

In order to prove the main result, given a boolean formula $\varphi(x_1,...,x_n)$
and an assignment S: $\{x_1, ..., x_n\} \rightarrow \{0,1\}$, we define:

$$\varphi_k = \varphi(S(x_1), ..., S(x_n))$$

where k, $0 \leqslant k \leqslant 2^n\text{-}1$, is the number whose binary representation is
the string $S(x_1) S(x_2)... S(x_n)$. We associate to the boolean formula φ
the generating polynomial p of the numbers φ_k, i.e.

$$p_p(t) = \sum_{k=o}^{2^n\text{-}1} \varphi_k t^h$$

Now, we can prove:

Theorem 6.1. Let $\varphi(x_1, ..., x_n)$ be a boolean formula containing m
operation symbols $b_j \in \{\wedge , \vee , \sim\}$. Then the generating polynomial $p\varphi$
can be generated by a slp of length 0 (n+m).

Proof — By induction on the construction of φ.

a) Let $\varphi = x_j$. Then

$$p_{x_j} = \sum x^k = (1+x) ... (1+x^{2^{j-1}}) .x^{2^j}.(1+x^{2^{j+1}}) ... (1+x^{2^n\text{-}1})$$

Let ζ be the constant boolean function $\zeta (x_1, ..., x_n) = 1$. Then the
corresponding generating polynomial is:

$$P_1 = \sum_{k=o}^{2^n-1} x = (1+x) . (1+x^2) ... (1 +x^{2^n\text{-}1}).$$

Moreover, we can compute all the polynomials P_ζ, P_1, ..., P_{x_n} in
0(n) arithmetic operations by means of:

$$P_{x_i} = \frac{x^{2^j}}{1+x^{2^j}} P_\zeta$$

b) The proof can be easily concluded by observing that

$$p_{\beta \wedge \zeta} = p_\beta \otimes p_\zeta$$
$$p_{\beta \vee \zeta} = p_\beta + p_\zeta - p_\beta \otimes p_\zeta$$

$p \sim_\beta = p_\zeta \cdot p_\beta$

The above formulas introduce $0(1)$ arithmetic operations for every $b_j \in \{\sim, \vee, \wedge\}$ in φ.

Now, let $\psi = Q_1 x_1 \ldots Q_n x_n \, \varphi(x_1, \ldots, x_n)$, where $Q_j \in \{\forall, \exists\}$, be a QBF, and $\#\psi$, as in sect. 3, the number of assignment trees accepting with respect to ψ; the following fact holds:

$$\#\psi = z_1 \ldots z_n \, \varphi(x_1, \ldots, x_n)$$

where

$$z_j := \text{if } Q_j = \forall \text{ then } \prod_{x_j \in \{0,1\}} \text{ else } \sum_{x_j \in \{0,1\}}$$

As a consequence, we can prove:

Theorem 6.2. $\#\psi$ can be computed by means of a slp $\prod$ of length $0(n)$.

Proof — The slp $\prod$ can be obtained as follows:

$$p0 := p_\varphi \; (= \sum_k \varphi_k x^k)$$

Comment: p_φ can be built up in $0(n+m)$ instructions, as in Th. 6.1.

If $Q_{n-j+1} = \forall$, then pj := $\boxed{2}$ $(p(j\text{-}1) \otimes (|p(j\text{-}1)))$
else pj := $\boxed{2}$ $(p(j\text{-}1) + (|p(j\text{-}1)))$

Comment: if $p(x) = p_0 + p_1 x + p_2 x^2 + \ldots$, then
$\boxed{2}$ $(p \otimes (|p)) = p_0 p_1 + p_2 p_3 x + p_4 p_5 x^2 + \ldots$
$\boxed{2}$ $(p + (|p)) = (p_0 + p_1) + (p_2 + p_3) x + (p_4 + p_5) x^2 + \ldots$
The output pn is the constant polynomial $p_n = \#\psi$.

As a consequence of Th. 6.1 and Th. 6.2, it is possible to compute the constant polynomial $\#\psi$ by means of a slp $\prod$ of length $0(|\psi|)$, where $|\psi|$ is the length of the formula ψ. Hence, by Th. 5.3, $\#\psi$ can be obtained with $0(|\psi|^2)$ arithmetic operations. We summarize these results in the following:

Theorem 6.3. (Main theorem) — The function #, which associates to every quantified boolean formula ψ the number of accepting assignment trees is computable by a RAM $(+, \dot{-}, *, \div)$ in $0\ (|\psi|^2)$ arithmetic operations.

Being # QBF complete in #P—SPACE, as proved by Th. 3.1, it follows easily:

Corollary 6.4. Every enumeration problem in the class #P—SPACE can be solved in polynomial time on a RAM $(+, \dot{-}, *, \div)$.

The converse also holds:

Theorem 6.5. For every function $f: \Sigma^* \longrightarrow N$ computable in polynomial time $T(|x|)$ on a RAM$(+, \dot{-}, *, \div)$ there exists a CTM M working in space less than $c.T^4\ (|x|)$ whose associated enumeration problem is solved by f.

Proof — First of all, we observe that the following bounds hold for f:

$$0 \leqslant f(x) \leqslant 2^{2^{T(|x|)}}$$

The CTM M works as follows:

1) on input x, M prints non deterministically the binary representation of every κ such that $0 \leqslant k \leqslant 2^{T(|x|)}$. Then, M enters a distinguished state q_1; the current configuration is thus $<q_1, x, k>$.
   (SPACE $< T(|x|)$)

2) from the configuration $<q_1, x, k>$, M computes deterministically the k-th bit in the binary representation of f(x), let c_k, and enters a distinguished state q_2. The current configuration is thus $<q_2, k, c_k>$.

   (SPACE $\leqslant T^4\ (|x|)$)
   To show that this is possible in space $T^4\ (|x|)$, we remember that:

   a) any RAM R can be simulated by a vector machine V [5] in time $O(T^2\ (|x|))$; so, it is easy to prove that there exists a vector machine V' that, on input $<x, k>$, computes c_k in time $O(T^2(|x|))$.

($|x|$)).

b) The vector machine V' can be simulated by a Turing machine working in space less than $c.T^4$ ($|x|$)) [5].

3) from the configuration $< q_2, k, c_k >$, M does the following moves:

if $c_k = 0$, then M enters a final non-accepting state q_N;

if $c_k \neq 0$, then M enters a distinguished state q_3, erases c_k, and non deterministically executes the following moves, using (eventually) the distinguished states q_4 and q_Y:

let $< q, j >$ be the configuration; then:

if $j \neq 0$, $< q_3, j > \rightarrow <q_3, j\text{-}1 >$ or $< q_4, j\text{-}1 >$
                 $< q_4, j > \rightarrow < q_3, j\text{-}1>$ or $< q_4, j\text{-}1 >$
if $j = 0$, $< q_3, j > \rightarrow < q_Y >$
                 $< q_4, j > \rightarrow < q_Y>$

where q_Y is a final accepting state.

(SPACE $\leq \log_2 k \leq T(|x|)$).
So, for every $c_k \neq 0$, M admits exactly 2^k accepting computations.
Therefore, the CTM M has exactly $f(x) = \sum_{k=o}^{2^{T(|x|)}} c_k . 2^k$ accepting computations, and works in space less than $c.T^4 (|x|)$.

Now, remembering that a VM $M \in V_k$ working in polynomial time can be simulated by a TM working in polynomial space [5] and that such a TM, as a consequence of Th. 6.3, can be simulated by a RAM $(+, \div, *, \div)$ working in polynomial time, we can conclude that every polynomial time VM can be polynomially simulated by a polynomial time RAM $(+, \div, *, \div)$. A straightforward padding argument allows to generalize the result, so obtaining:

Theorem 6.6. $V_k \xrightarrow{p} $ RAM $(+, \div, *, \div)$.

Then, we can complete the diagram shown in section 3 as follows:

$$V_k \xleftrightarrow{\ P\ } RAM(+, \div, *,) \xleftrightarrow{\ P\ } RAM(+, * \wedge)$$

$$\Big\uparrow \qquad\qquad \Big\uparrow \qquad\qquad \Big\uparrow$$

$$RAM(+, \div, *, \div) \xleftrightarrow{\ P\ } RAM(+, \div, *, \downarrow) \xleftrightarrow{\ P\ } RAM(+, *, \downarrow)$$

7. Equivalence problem for numbers succinctly represented

A slinghtly different version of Th. 6.5 is the following:

Theorem 7.1. For every RAM $(+, \div, *, \div)$ R, which computes in polynomial time a function f: $\Sigma^* \longrightarrow N$, there exists a deterministic Turing machine M_R which accepts in polynomial time an input $x \in \Sigma^*$ and prints a slp on the structure $\mathscr{N}$, $\varphi(x)$, with $v(\varphi(x)) = f(x)$.

Proof — Let R be a RAM computing the function f; by Th. 6.5, we can construct a Turing machine M_R working in polynomial time with associated enumeration function f. Since #QBF is complete in #P—SPACE, we can associate, in polynomial time, to every $x \in \Sigma^*$, a quantified boolean formula $\psi_{M_R}(x)$ such that:

$$\#\psi_{M_R}(x) = f(x)$$

Furthermore, by Th. 5.2, it is possible to associate with $\psi_{M_R}(x)$ a straigth line program $\varsigma_{M_R}(x)$, such that:

$$\psi_{M_R}(x) = \varsigma_{M_R}(2)$$

(we remark that $\varsigma_{M_R}(x)$ is a polynomial of degree 0).

Finally, by Th. 4, we can construct in polynomial time a straigth line program on $\mathscr{N}$, $\Phi(x)$, such that:

$$v(\Phi(x)) = \varsigma_{M_R}(2) = \#\psi_{M_R}(x) = f(x)$$

The meaning of this theorem is that, by using a particular representation of numbers as slp's, it is possible to simulate a RAM($+, \div, *, \div$), working in polynomial time, by means of a polynomial time deterministic Turing machine.

This is a rather surprising result; since we have proved that polynomial time on RAM's is equivalent to polynomial space on TM's, we can suspect that the usual relations of equality and order on numbers represented as slp are difficult to compute. To show that, let us consider the following problem:

EQUIVALENCE PROBLEM FOR SLP's (EPSLP)

INSTANCE: A pair $< P_1, P_2 >$ of slp on $\mathcal{N}$.
QUESTION: Is it $v(P_1) = v(P_2)$?

We can prove:

Theorem 7.2. EPSLP is P–SPACE complete.

Proof – First, we prove that EPSLP P–SPACE. In fact, we can easily construct a RAM($+, \div, *, \div$) that, on input $< P_1, P_2 >$, writes $v(P_1)$ in a register and $v(P_2)$ in another one (the time needed is $0(|P_1|+|P_2|)$) and then verifies whether the contents of these registers are equal (in time $0(1)$). Hence, EPSLP can be solved in polynomial time on a RAM ($+, \div, *, \div$) and, by Th. 6.5, is in P–SPACE. In order to prove that EPSLP is P–SPACE hard, we observe that it is possible to associate with every QBF ψ, in polynomial time, a slp P_ψ such that $v(P_\psi)$ is the number of assignment trees satisfying ψ (see Th. 7.1). It follows that:

$$v(P_\psi) \neq 0 \text{ iff } \psi \text{ satisfiable.}$$

Since the problem "SATISFIABILITY OF QBF" is P–SPACE complete [10], it turns out that EPSLP is P–SPACE hard, even in the particular case of $P_2 = 0$.

A similar problem was studied by Schönage [6], for the particular case of slp's using only the operations $+$ and $*$. In this case, the equivalence problem turns out to be co-NP.

Our result gives a strong evidence to the Simon's conjecture on the power of the integer division with respect to the other arithmetical operations [9].

References

[1] Bertoni A., Mauri G., Sabadini N., A characterization of the class of functions computable in polynomial time by Random Access Machines, Proc. 13th ACM STOC, 1981, 168−176

[2] Garey, M.R., Johnson D.S., Computers and intractability, W.H. Freeman and Co. San Francisco, California, 1979

[3] Hartmanis J., Simon J., On the power of multiplication in Random Access Machines, 15th IEEE Symp. on Switching and Automata Theory, 1974, 13-23

[4] Percus J.K., Combinatorial Methods, Springer, Berlin, 1971

[5] Pratt V.R., Stockmeyer L.J., A characterization of the power of vector machines, JCSS 12, 1976, 198-221

[6] Schönage A., On the power of Random Access Machines, Proc. 6th ICALP, Lect. Not. in Comp. Sci. 71, Springer, Berlin, 1979, 520-529

[7] Simon J., On some central problems in computational complexity, Ph. D. Thesis, Dept. of Comp. Sci., Cornell University, Ithaca, N.Y., 1975

[8] Simon J., On the difference between the one and the many, Lect. Not. in Comp. Sci. 52, Springer, Berlin, 1977, 480-491

[9] Simon J., Division in good, Comp. Sci. Dept., Pennsylvania State University, 1979

[10] Stockmeyer L., Meyer A., Word problems requiring exponential time, Proc. 5th ACM Symp. on Theory of Computing, 1973

[11] Valiant L.G., The complexity of enumeration and reliability problems, Res. Rep. CSR−15−77, Dept. of Comp. Sci., Univ. of Edinburgh, 1977

[12] Valiant L.G., The complexity of computing the permanent, Theoretical Computer Science 8, 1979, 189-202

ACKNOWLEDGEMENTS

This paper has been supported by Ministero della Pubblica Istruzione, in the frame of the project "Theory of algorithms"

Annals of Discrete Mathematics 25 (1985) 91-108
© Elsevier Science Publishers B.V. (North-Holland)

A REALISTIC APPROACH TO VLSI RELATIONAL DATA-BASE PROCESSING*

M. A. BONUCCELLI, E. LODI, F. LUCCIO,
P. MAESTRINI, L. PAGLI

Dipartimento di Informatica Università di Pisa

We propose new algorithms for VLSI mesh of trees structures to store and to process relational data bases. It is observed that some previous approaches to this problem suffer from the limitation that a relation might not be entirely contained in a single VLSI chip. In this new proposal, relations of *arbitrary size* can be handled by means of several chips of *fixed size*, still retaining good time performances. It is shown that the time needed to perform elementary and relational operations is *logarithmic* in the chip size and *linear* in the number of chips needed to store a relation, and in the bit length of the relation elements.

1. Introduction

The proposal for highly dedicated parallel systems, based on VLSI technology, are becoming more and more realistic. In fact the number of gates that can be integrated in a single chip is steadily increasing, and the processing capability of the chip is greater. [1]

In addition, a problem which has been only recently approached, is the one of handling a set of data of size greater than that of the chip [2, 6].

This problem is relevant because most of the proposed VLSI systems can be applied only to bounded size data, so severely restricting the field of their applicability.

In particular we will refer to VLSI design of a data base machine

* This work has been supported by Ministero della Pubblica Istruzione under a research grant.

suitable for the relational model. Several proposals have been made on this matter, which substantially differ one from the other with respect to connection pattern of the processing elements, i.e. : systolic array [3], binary tree [4,5], and mesh of trees [6].

A major practical problem is the allocation of relations which do not fit in a single chip. This problem is solved only for the scheme based on binary trees [7], while the definition of the operations to be performed on the relations is disregarded in every scheme. On the other hand, it is important for a data base machine to be able to handle relations of any size, independently on the size of the chips, as was pointed out in [8].

In this paper we propose an extension of the data base machine based on VLSI mesh of trees, which solves several practical problems left open in the original paper [6].

Among the VLSI data base machines proposed so far for relational data bases, the one based on mesh of trees chip allows to perform all the standard operations on data with the least time complexity.

A qxq mesh of trees is a set of q^2 nodes arranged as a qxq array. The nodes in the i-th row (i = 1, . . . , q) are connected by a complete binary tree, as well as the nodes in j-th column. The structure can be laid out in Θ (q^2 $\log^2$ q) area [9]. By the assumption that every relation of the data base can be contained in a single chip and the I/O bandwidth is not smaller than the relation size, even the most complex relational operation can be performed in $O(\log n)$ time units, if n is the relation size [6].

Another assumption generally adopted is to put the array of data on a rectangular chip, so having a number of rows equal to the number of records of a relation, and a number of columns equal to the number of elements of each record. In almost all real cases, however, the number of records is much greater than the number of elements of each record, hence the resulting chip would have drastically different sides, against the current manufacturing criteria that lead to chips that are almost square.

To overcome the above problems we analyze how to arrange a rectangular relation in a square chip, and then how to partition a large relation into different chips. In both cases the algorithms for a set of classical relational database operations are derived, and their time complexity is analyzed.

Finally the two extensions are meshed together; the result obtained is the definition of a data base machine able of handling relations of any size, and which still guarantees a good time performance.

The operations described in this paper can be divided in two groups, namely, *elementary operations,* (I/O and simple search-update operations) and *relational operations* (Union, Set difference, Cartesian product, Projection, Selection, Join). This set of operations is simple but complete [10]. It must be seen merely as a working proposal, since different sets of operations could be easily defined on the chips, with similar techniques.

2. Chip organization

A relation A consists of m h-tuples $a_1, \ldots, a_m$. Every h-tuple a_i $(1 \leqslant i \leqslant m)$ is formed by h elements $a_{i1}, \ldots, a_{ih}$. The elements a_{1j}, $a_{2j}, \ldots, a_{mj}$ $(1 \leqslant j \leqslant h)$ are drawn from the same domain of values. Without loss of generality we assume that all the elements are b bits long.

A qxq mesh of trees (MT) is composed by rows $f_1, \ldots, f_q$ and columns $\gamma_1, \ldots, \gamma_q$; the leaves $\lambda_{11}, \lambda_{12}, \ldots, \lambda_{qq}$ lie at the intersection of rows and columns (λ_{ij} lies in row f_i and column γ_j). The 2q row and column trees are denoted by $rt_1, \ldots, rt_q$ and $ct_1, \ldots,$ ct_q respectively.

The input and output of data and messages in an MT is done through the roots of row and column trees. MT's can be used to store relations or to perform operations on elements coming from an outside memory. To fully exploit the potential processing parallelism of an MT, 2q units of data must be input and output in parallel from it. The memory devices storing the relations must have an I/O bandwidth of 2q units of data per clock time. MT's can be conveniently used as memory devices with the required I/O bandwidth. When used as a memory, an MT needs a more complex structure: in the following we refer to Storage Mesh of Trees (SMT), and to Processing Mesh of Trees (PMT), according to the use. To simplify chip design, we will assign the same basic structure to SMT and PMT nodes. Since each node is an elementary processor, what will distinguish between SMT and PMT is the set of microprograms loaded into the control unit of any single node. A node must be able to hold data, and to send data to its father and sons. Flows of data up and down the tree can take place contemporarily in the same node. For this purposes each node has two registers that can be used for permanent and/or temporary storage of data elements. In addition, an Availability Flag register (AF) will be used to indicate whether the leaves that can be reached from the node are storing tuple elements or not. The registers are connected to the ones of other nodes via bidirectional links, to implement the connection pattern of meshes of trees. A node will be also called to perform some elementary arithmetic and logic operations.

The architecture of a possible computing system that uses SMT's and PMT's is shown in fig. 2. A control processor supervises and coordinates the processing of the MT's by issuing commands and instructions. Besides, it supervises the flow of data of the system. Mass memory devices are present in the system to store data not involved in the processing.

3. Use of square mesh of trees for rectangular relations

We now discuss how an mxh relation A (m $\gg$ h), can be stored in a qxq mesh of trees in case of q $\geqslant \sqrt{m\,h}$. When q $< \sqrt{m\,h}$, more than one meshes of trees are needed to hold the relation; such situation will be discussed in the following sections.

A relation A can be efficiently handled in a square MT if q is the smallest power of 2, $q = 2^k$, such that $2^k \geqslant \sqrt{m\,h}$. A is arranged in the following way: tuples a_1, a_{q+1}, a_{2q+1}, . . . are stored in the first row. Tuples a_2, a_{q+2}, a_{2q+2}, ... are stored in the second row, and so on. If $2^{i-1} < h \leqslant 2^i$, for each tuple 2^i-h colums at its right are left unused, so that each row subtree with the root at level i, (leaves are at level 0) has exactly one tuple stored in its leaves. The set of such leaves is called "tuple subrow"; the root at level i is called "tuple root", and the subtree having such a root is called "tuple subtree" (fig. 3). For example, the columns $\gamma_{h+1}, \ldots, \gamma_{2}i$ will be reserved for $a_1, \ldots, a_q$, but never used. Every row stores $x = \lfloor q/2^i \rfloor$ tuples. Under the assumption that $q = 2^k$, at most $x = 2^{k-i}$ tuples are stored in each row. It can be easily shown that, in the worst case, at most half of the MT area is left unused.

The Availability Flag register of each tuple root is set to 1 if the leaves under it store an h tuple, and is set to 0 if the leaves contain meaningless information (e.g., a tuple to be cancelled).

INSERT an h-tuple in relation A

Input. An h-tuple $a_i = a_{i_1}, \ldots, a_{ih}$, and the address of the j-th subrow of a given row β_y.

Result. a_i is stored in a subrow, $AF = 1$ in the corresponding tuple root.

Assumption. The control processor knows the availability of tuple roots, and decides the subrow that will store the tuple.

Procedure. The tuple $a_{i_1}, \ldots, a_{ih}$ to be inserted is sent to the root of the row tree rt_y, with each element followed by its column index. The pairs a_{ik}, k are sent in order of increasing k. The element $a_{i_1}, 1$ is preceded by an "insert" command containing the relative address $j-1$ of the j−th destination subrow. This address is used by the internal nodes of the row tree, at the levels above the tuple root, to select the path from the root of the tree to the tuple root. (Such a path can be algorithmically determined from the binary representation of $j-1$. See for example [6].) Similarly, each element a_{ik} is steered in the subtree, towards the k−th leaf of the tuple subrow.

The time required by INSERT is of order $h + \log_2 q$. In fact $\log_2 q + 1$ time units are required to store a_{i_1}, and additional $h-1$ time units to store the other elements in a pipeline fashion.

A simple extension of the algorithm allows to store a whole relation A in an SMT. The algorithm is applied in parallel to all the rows to store, in a pipeline fashion, the tuples a_i, a_{q+i}, ... in each row $\int_i$. The time required is $2^{k-i} (h + \log_2 q) = x(h+k)$.

Another important operation is:

SEARCH a tuple in relation A.

Input. An h'-tuple a_i, $h' \leqslant h$.

Result. The number of times a_i is present in every row, stored in the root of the rt.

Procedure. The pairs a_{ij}, j are input into the trees rt_j. Each internal node of rt_j broadcasts this information to the leaves. All the leaves λ_{iy} such that $y \equiv j (\mod 2^i)$ send the received pair a_{ij}, j upwards, to the root of the column trees. Each pair is then broadcast again from the root to all the leaves. After $3\log_2 q$ time units, each leaf λ_{xy} receives the proper pair a_{ij}, j with $y \equiv j \pmod{2^i}$, and $1 \leqslant x \leqslant q$, (see fig. 4). Here the received element a_{ij} is compared with the resident element. If the elements match (or do not match), a bit 1 (or 0) is sent up the row tree. The 2^i-h rightmost leaves of each tuple subrow, that do not contain elements, send a 1. The two bits received from the sons are AND-ed in each node, and the result sent upwards, until the tuple root is reached, where the two bits are AND-ed with AF. Here, a 1 (or a 0) indicates matching (or not matching) of the whole tuple. From this level on, nodes add values received from the sons and send the results upwards.

After $\log_2 q$ additional time units, the roots of row trees will compute the desired results.

Hence the operation can be carried out in $4\log_2 q$ time units.

SEARCH can be simply modified to delete, or to remove the duplicates of a given tuple.

The last important elementary operation is the output operation.

OUTPUT a tuple from a relation A

Input. The address of the j-th subrow of a given row $\mathcal{f}_y$, from which the tuple is to be extracted. An ordered sequence $\pi = p_1, \ldots, p_{h'}$, of a subset of the integers $1, \ldots, h$ ($h' \leqslant h$).

Result. The ordered sequence $a_{ip_1}, \ldots, a_{ip_{h'}}$, of h' elements of the h-tuple a_i available at the root of rt_y, in h' consecutive clock times.

Procedure. j-1 followed by the sequence π is sent to the root of rt_y. π is routed to the j-th tuple root (see the operation INSERT). In the tuple subtree j, each element p_i is sent to the p_i-th leaf. When the leaf receives this value, it sends its resident element upwards. Therefore, the elements $a_{ip_1}, \ldots, a_{ip_h}$, reach the root of rt_y in a compact sequence.

The algorithm requires $2\log q + 1$ time units, to make the first element a_{ip_1} available at the root of rt_y. The remaining elements are output in sequence in h'-1 additional time units.

The *output of whole relation* can be performed by issuing in parallel an OUTPUT command to every row, to output the leftmost h-tuple contained in the row. This causes the output (in parallel) of the tuples $a_1, \ldots, a_q$, in $2\log q + h$ time units.

The operation is repeated x-1 times, to output $a_{q+1}, \ldots, a_{2q}$; then $a_{2q+1}, \ldots, a_{3q}$ etc., each group in h time units.

Total time is $2\log q + xh$.

The processing of relational operations is carried out in a square PMT. If a large PMT is available, namely, one whose size is nxn, with

$n \geqslant m$, then these operations can be directly performed as already shown in [6]. In a more realistic assumption, however, the PMT size is equal to that of the SMT, i.e. qxq. In this case the relational operations must be performed in the PMT on blocks of q tuples each time. This approach will be followed also when the relation does not fit in a unique SMT, as discussed in the following section. Hence we postpone the description of the relational operations, after the second memorization scheme is discussed.

4. Storing a relation in more than one SMT's

In this section we discuss how relations that do not fit in one qxq SMT can be processed.We also assume that qxq is the size of PMT.

We restrict our study to the case where only one h-tuple is stored in an SMT row, and $q < m$. A relation A is therefore stored in p SMT's, $pq \geqslant m$. The simple extension to the general case will be discussed in the last section.

The elementary operations of insertion, search and output can be easily performed in parallel on the p SMT's storing A, as immediate extensions of [6]. Note that to send the inputs of a given operation in parallel to all the pq rows, such inputs must be multiplicated by using a PMT [9]. For this purpose, 2k logq time units are needed to generate the desired copies of the inputs, where $(k-1) q \leqslant p \leqslant kq$.

We consider now the complete set of relational operations described in [10], namely, *union, set difference, cartesian product, projection* and *selection*. In addition we discuss a direct realization of the *join* operation, which is useful to solve some particular relational query.

Many of relational operations, in this scheme, require sorting as preprocessing. Sorting is therefore discussed. We will see how, in this case, the time complexity of the operations can be drastically reduced.

The sorting algorithm below is a brief repetition of the one presented in [2]. By *block B_i* we indicate the data stored in SMT_i.

SORT the tuples of a relation A

A is stored in p SMT's. To define sorting of tuples, these are treated as strings, assigning a priority among the columns, with the element of higher priority as the most significant of the string, and assigning a lexicographic order relation among elements.

Step 1. Each block is separately sorted using the standard algorithm of [9].

Step 2. Block pairs B_i, B_j are compared in the PMT to produce two new blocks B'_i and B'_j containing the q smaller tuples and the q greater tuples respectively.

The selection of the block pairs to be compared is done in a merge sort fashion, considering blocks as the smallest units of data.

The total time needed to sort a relation stored in p SMT's is of $0((h+\log q)\ p\log p)$.

In the following, we assume that each relation contains distinct tuples.

The result of a relational operation is overwritten onto the input relations; however, these relations can be saved by previous duplication, that can be obviously done directly between SMT's in $\log q + h$ time units (independently of relation size).

4.1 Complete Set of Relational Operations.

UNION of two relations A and B (denoted by AUB)

The set union of A and B considered as sets of tuples is performed only on relations having the same number h of tuple elements.

Input. Two relations A and B of size $m_A \times h$ and $m_B \times h$, stored in p_A and p_B SMT's respectively.

Result. Relation C = AUB, stored in the p_A SMT's of A, and in part of the p_B SMT's of B (if needed).

Preprocessing. SORT A and SORT B, according to the same specified priority among columns.

Procedure. Consider A and B as partitioned in sorted blocks A_i, $1 \leqslant \leqslant i \leqslant p_A$, B_k, $1 \leqslant k \leqslant p_B$ each contained in an SMT. Starting from A_1 and B_1, pairs of blocks are output from SMT's and input into the PMT. Blocks from relation A are input in the row trees, and those of B in the column trees.

Here blocks A_i and B_j are compared tuple-wise in $h + 2\log q$ steps. If two elements match, a 0 flag is generated in the leaf. These flags are sent up the tree, OR-ed on the nodes, such that a 0 emerges in the column tree root, if two tuples match. The flag values so obtained are sent to the SMT storing B_j, where they are AND-ed with the Availability Flag in the roots of the row trees to delete tuples of B_j already present in A.

If the last element of A_i is smaller than that of B_j, (such information can be easily deduced during the above comparison step), A_{i+1} and B_j will be successively compared; otherwise A_i and B_{j+1} are compared. Therefore the procedure requires at most $p_A + p_B - 1$ comparisons between blocks.

The overall time required by the procedure is $0((p_A + p_B)(h + \log q))$. (Time required by the sorting preprocessing is not included).

The second relational operation is the set difference.

DIFFERENCE of two relations A and B (denoted by A−B). C=A−B consists of all the tuples in A not present in B. This operation can be performed only on relations with same h.

Input. Two relations A and B of size $m_A \times h$ and $m_B \times h$, stored in p_A and p_B SMT's, respectively.

Result. An $m_C \times h$ relation C=A−B stored in the SMT's of A.

Preprocessing. SORT A and SORT B.

Procedure. As in the algorithm for UNION, A and B are suddivided in blocks, that are sent pairwise, A_i, B_k, to the PMT, to find matching tuples. The result of comparison is used to eliminate matching tuples from A_i, with the same strategy of the UNION.

The overall time complexity is the same of UNION.

PRODUCT of relations A and B

This operation yield to the cartesian product of A and B treated as sets of tuples.

Input. Two relations A and B, of size $m_A x h_A$, and $m_B x h_B$, stored in p_A and p_B SMT's, respectively.

Result. A relation C=AxB of size $(m_A m_B) \cdot (h_A + h_B)$, where in each tuple the first h_A elements constitute a tuple of A, and the following h_B elements constitute a tuple of B.

Procedure.

Step 1. m_B copies of each tuple of A are generated. In fact, starting with a_1, tuples are sent to the PMT, to be reproduced in q copies. This operation must be repeated $\lceil m_B/q \rceil$ times per tuple. Each group of q copies is transferred from PMT to the SMT's reserved for C, first h_A columns, without leaving empty rows. Tuple a_i is stored in the y-th row of the x-th SMT of C, where:

$$x = \begin{cases} \lceil (i-1)m_B/q \rceil & \text{if } x \not\equiv 0 \ (\text{mod } q) \\ ((i-1)m_B/q)+1 & \text{otherwise} \end{cases}$$

and

$$y = \begin{cases} 1 & \text{if } x \equiv 0 \pmod{q} \\ ((i-1)(m_B - \lfloor m_B/q \rfloor q)+1) \pmod{q} & \text{otherwise.} \end{cases}$$

These addresses are assigned to the tuples by the control processor.

Step 2. Relation B is output in parallel from its SMT's, in m_A successive times, and input in the columns from $h_A + 1$ to h_B of SMT's storing C. Each time t_i, B is steered to the rows storing a_i, according to the addresses computed in step 1.

The overall time is $0(m_A^2/q+\log q)$.

PROJECTION of a relation A over k domains

The projection of A over a specified subset of $k \leqslant h$ columns, is the set of the k-subtuples of A, that are pairwise different in at least one column.

Input. A relation A, stored in p SMT's, of mxh elements and a subset $c_1, \ldots, c_k$, $k \leqslant h$ of column indices.

Result. An m'xk relation, whose tuples are stored in the SMT's of A in the leaves of the row trees with AF = 1 at the root.

Preprocessing. SORT A over the columns $c_1, \ldots, c_k$.

Procedure. After preprocessing, the tuples having the same values in $c_1, \ldots, c_k$ are contiguous, and they may lay on different SMT's. Each SMT executes a standard projection [6], to suppress duplicate tuples locally. SMT contents are then compared in PMT in pairs SMT_i, SMT_j, starting with i=1 and j=2.

If SMT_i, SMT_j contain a matching tuple, this is deleted from SMT_j. Two cases then discriminate the following steps. If SMT_j has become empty (i.e., AF=0 for all row tree roots), then the process is repeated with $j=j+1$. If SMT_j is not empty, the process is repeated with $i=j$ and $j=j+1$. The procedure ends when $j>p$, that is after $p-1$ SMT comparisons.

The total time (except for the preprocessing) is of $O(plogq)$. In fact, each of the $p-1$ comparison steps requires $O(logq)$ time for SMT comparison, and $O(logq)$ time to test emptiness of SMT_j.

The next relational operation, called SELECTION, makes use of Boolean expressions that are used as conditions to select the subset of tuples of a given relation A, satisfying such a condition. We assume that the Boolean expressions are in disjunctive normal form, that is $C_1 \lor C_2 \lor \ldots \lor C_r$, $r \geq 1$, where each C_i is called a *clause*. Clauses are of the form $b_1 \land b_2 \land \ldots \land b_s$, $s \geq 1$. Each b_i is a Boolean variable of the form x_1 op x_2, where x_1 represents a domain; x_2 is either an element or a costant; op is a comparison operator of the set $\{<, >, =, \neq, \leq, \geq\}$. b_i is true if and only if x_1 op x_2 is verified.

Note that it is not necessary to use the Boolean operator $\neg$ (not), since any Boolean variable containing it can be restated by using the complement of op. For istance, $\neg(x_1 \geq x_2)$ can be transformed into $(x_1 < x_2)$.

The operation is as follows:

SELECTION of a subset of tuples satisfying a specific condition

Input. A relation A, and a Boolean expression, $C_1 \lor C_2 \lor \ldots \lor C_r$. The Boolean variables appearing in every C_i are built with the domain values of A.

Result. A susbset of tuples of A that verifies the Boolean expression, detected by AF=1 in the roots of the SMT's row trees.

Procedure. This operation can be performed by the p SMT's storing relations. PMT is used only to generate p copies of the Boolean expression, one for each SMT.

Clauses C_i are sent to the SMT's, one at time. If, in a Boolean variable $b_i = x_1$ op x_2, x_2 is a constant, such a variable is input in parallel in every row tree, and duplicated as in the algorithm SEARCH. When the variables reach the leaves, the comparison specified by "op" is performed, and a one bit result is issued. The result is sent up to the root of the row tree, and here AND-ed with the result of the previous evaluations of Boolean variables of the same clause.

Note that in this case, the Boolean variables of a clause can be evaluated in pipeline with an algorithm similar to that of SEARCH.

The operation is more complex if a clause contains $b_i = x_1$ op x_2, with both x_1 and x_2 domains, or if two Boolean variables refer to the same domain. If x_2 is a domain, b_i must be evaluated in two phases. In the first phase, x_2 is input in each row tree and the corresponding element value is sent up to the root; secondly, such a value is used like a constant, and b_i can be evaluated as before. If two Boolean variables refer to the same domain, they must be evaluated sequentially.

We assume that the clauses are preprocessed by the control processor and sent to the SMT's in two or more parts, so that each part can be evaluated in pipeline without the above mentioned conflicts. The results of the evaluation of clauses are composed at the root of the row trees by OR-ing them.

The time complexity for SELECTION depends on the number of Boolean variables and on the number of parts in which each clause must be divided. Each part is evaluated in pipeline in $2\log q + t - 1$ time units, where t is the number of Boolean variables. Therefore the overall time complexity is of $O(y\log q + T)$, where T is the total number of Boolean variables in the expression, and y is the number of parts in which the expression is divided. Therefore, the time complexity for selection depends on the Boolean expression only. The previous analysis does not take into account the time needed to generate the proper number of copies of the Boolean expression at the input. The time is negligible for $p < q$, while it becomes of $O(p/q\log q)$ for $p \geqslant q$ as discussed at the beginning of this section.

We now show how to perform the operation of JOIN. Although

JOIN can be performed by a suitable sequence of the relational operations previously defined, it is useful to have a fast algorithm for it.

Consider two relations $A, m_A x h_A$ and B, $m_B x h_B$ and a set $c_1, \ldots, c_h$ of columns common to A and B.

In the cartesian product AxB, columns $c_1, \ldots, c_h$ are repeated twice. Call them c_{i_A} and c_{i_B}, to indicate their origin. We recall that JOIN of A and B is a relation consisting of the tuples of AxB whose elements coincide c_{i_A} and c_{i_B}, for all i, and where columns c_{i_B} are suppressed.

JOIN of relations A and B over $c_1, \ldots, c_h$

Input. Two relations A and B of size $m_A x h_A$ and $m_B x h_B$, and a subset of columns $c_1, \ldots, c_h$ common to A and B.

Result. A relation C of size $m_c x (h_A + h_B - h)$.

Preprocessing. PROJECTION of A and B over $c_1, \ldots, c_h$. The algorithm for the projection is enabled to count, for each resulting tuple a_i (or b_i) the number of original tuples n_{a_i} (or n_{b_i}) matching with a_i (or b_i) in $c_1, \ldots, c_h$. n_{a_i} or n_{b_i} are computed at the row tree roots, as the number of occurrences of a searched tuple was computed in SEARCH.

Procedure.

Step 1. Compare A and B to get the number of occurrences of each tuple of A and B in the resulting relation C, in the following way.

Let A be divided in blocks $A_1, \ldots, A_{p_A}$, and B in blocks $B_1, \ldots, B_{p_B}$. Starting from A_1 and B_1, one block of A and one of B are input in the PMT (A blocks in row trees, B blocks in column trees).

In the leaves of PMT the elements of tuples drawn from domains

$c_1, \ldots, c_h$ are pairwise compared. If two tuples a_i and b_j match, n_{a_i} and n_{b_j} are exchanged between them, along with their indices i and j. With a mechanism similar to that of SORT, the next two blocks to be compared are determined, until all blocks are compared.

Step 2. The positions of tuples of A and B in the result C are determined by computing $\sum_{i=1}^{j} n_{a_i}$ for each tuple a_j and $\sum_{i=1}^{j} n_{b_i}$ for each tuple b_j.

This can be done in the PMT by broadcasting such numbers to the leaves, and then summing up in each row only the numbers under the main diagonal of the mesh.

Step 3. Relation C is actually set up, by using a PMT as a generalized interconnection network [9].

The information obtained in the first and second steps in used to generate the proper number of copies of each tuple of C, and to send them in the right places of the SMT's that will store C. (For details on this operation, see a similar procedure PMT-OPERATION 4 of [6]).

The first step requires $0(p'(\log q + h))$ where $p' = \max(p_A, p_B)$, the second one requires $0((p_A + p_B)\log q)$ and the last again $0((p_A + p_B)\log q)$. Hence the overall time complexity (except for preprocessing of PROJECTION) of JOIN is $0((p_A + p_B)\log q)$.

5. Concluding remarks

In section 3 we have seen how to arrange a mxh relation in only one square qxq SMT.

In section 4 we considered the problem of processing relations stored in more than one SMT's, with just one tuple per row.

Let us now consider the most general case, that is a relation stored in more than one SMT's, with more than one tuple per row. Let x be the maximum number of tuples that can be stored in a single row of each SMT; we have xpq $\geq$ m.

It is easy to see that the elementary operations are a simple exten-sion of that seen in section 3; here the input of the operations to be performed must be reproduced in q copies by the PMT, which implies an increase of time of 2 $\lceil p/q \rceil$ logq.

The relational operations can be performed as in section 4. The only difference is that a block does not coincides with the contents of a whole SMT, rather with h columns. Therefore the time required increases by a factor of x, because q is now greater than $\lceil m/px \rceil$.

Although an exact comparison in time performance between diffe-rent proposed VLSI data base machines would require a much finer analysis, we can assert than the machine described in the present paper has two main advantages over the previous proposals: it is very fast and it can efficiently handle relations of any size.

In fact, the time needed to perform elementary and relational operations is, in general, of the order of the *logarithm* of the dimen-sions of the chip, and *linear* in the number of chips storing a relation, and in the bit length of the relation elements.

A further development of this proposal is to exactly define the overall architecture of the system; in particular an efficient intercon-nection between SMT's and PMT's is needed to fully exploit the parallelism of processing of the mesh of trees.

References

[1]   Computer: *"Special Issue on Highly Parallel Computing."* January 1982.

[2]   M. A. Bonuccelli, E. Lodi, L. Pagli, : *"External Sorting in VLSI".* Technical Report, S–83–10, Dipartimento di Informatica, Università di Pisa, March 1983. Also, to appear in IEEE Trans. on Computers.

[3]   H. T. Kung, P. L. Lehman: *"Systolic (VLSI) arrays for relational data base operations".* Proc. ACM–SIGMOD 1980 Int. Conf. on Management of Data, May 1980, pp. 105–116.

[4]   S. W. Song.: *"A highly concurrent tree machine for data base applications".* Proc. IEEE 1980 Int. Conf. on Parallel Processing, pp. 259–260.

[5]   M. A. Bonuccelli, E. Lodi, F. Luccio, P. Maestrini, L. Pagli, : *"A VLSI tree machine for relational data bases".* Proc. 10-th Annual Symp. on Computer Architecture, June 1983, pp. 67–73.

[6]   M. A. Bonuccelli, E. Lodi, F. Luccio, P. Maestrini, L. Pagli.: *"VLSI mesh of trees for data base processing".* Proc. CAAP 83, March 1983, to appear as Lecture notes in Computer Science 159, Springer Verlag, pp. 155–166.

[7]   J. L. Bentley, H. T. Kung.: *"A tree machine for searching problems".* proc. IEEE Int. Conference on Parallel Processing, Aug. 1979.

[8]   B. W. Arden, R. Ginosar. : *"A single-relation module for a data base machine".* proc. 8-th Annual Symp. on Computer Architecture, May 1981.

[9]   F. T. Leighton. *"New lower bound techniques for VLSI".* Proc 22nd Annual Symposium on Foundations of Computer Science, 1981, pp. 1-12.

[10]   J. D. Ullman: *"Principles of Database Systems",* Computer Science Press, Potomac (Md), 1980.

Annals of Discrete Mathematics 25 (1985) 109-124
© Elsevier Science Publishers B.V. (North-Holland)

ON COUNTING BECS

R. CASAS, J. DÍAZ, M. VERGÉS

Facultat d'Informatica. Universitat Politècnica de Barcelona.

In the first part of this article, we introduce an algebraic formalization for binary trees and their associate the *compatible linked forests (becs)*[2]. In the second part, we compute the number of *becs* with *n* vertices. Starting from the recursive generation of the *becs* and using generating function techniques we obtain an equation characterizing the number of *becs* of a fixed size.

1. Definitions

All through this work we refer only to *binary trees,* that from now on we shall shorten as trees. A *forest* will be any non empty set of different trees. We will consider only finite forests.

Formally, a forest will be a quatruple *(V, F, l, r)* determined by a finite set *V* of *vertices,* a subset $F \subset V$ of *leaves* and two one-to-one mappings

$$l, r : \ V - F \ \rightarrow \ V$$

such that to each *interior* vertex *v, l(v)* corresponds to the *left son* of *v,* and *r(v)* corresponds to the *right son* of *v.* For the standard notation about trees and graphs, see for ex. |3|.

Keywords: Binary trees, compatible linked forests *(becs),* linkage.

(1) The name bec *is an acronym for* bosc enllaçat compatible, *that is the catalan translation of compatible linked forest.*

Let us consider the following *strict* (i.e. antirreflexive) *order* relation on the vertices of a forest, defined by:

If v and v' are vertices then

$$v < v' \Leftrightarrow \exists\, n > 0 \,\, \exists\, v_0, \ldots, v_n \quad \left| \begin{array}{l} v \ = v_0 \\ v' = v_n \\ \forall\, i < n \quad v_i = l(v_{i+1}) \text{ or } v_i = r(v_{i+1}) \end{array} \right.$$

Definition 1. Given a forest, we say it is *compatible* when it does not contain any two trees such that one of them is equal to a subtree of the other.

Example of a compatible forest:

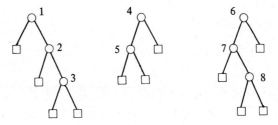

Definition 2. For any given forest we shall say that two vertices are equivalent if they are roots of isomorphic subtree. It is possible to characterize this equivalence by means of the following recursive definition: Let (V, F, l, r) be a forest; two vertices $u, v \in V$ are said *equivalent* $(u \equiv v)$, if

    a) either $u \in F$ and $v \in F$

    b) or $l(u) \equiv l(v)$ and $r(u) \equiv r(v)$

From now on, we shall denote by R this equivalence relation. In the above example, the equivalence classes are:

$$F, \{3, 5, 8\}, \{2, 7\}, \{1\}, \{4\}, \{6\}$$

Taking the quotient of this equivalence relation we could consider the two induced mappings $\tilde{l}$ and $\tilde{r}$, defined as those mappings that make commutative the following diagram:

$$
\begin{array}{ccc}
V-F & \xrightarrow{\ l,\ r\ } & V \\
\Big\downarrow & /\!/\!/ & \Big\downarrow \\
(V-F)/R & \xrightarrow{\ \tilde{l},\ \tilde{r}\ } & V/R
\end{array}
$$

In other words, if $[v]$ denotes the class of vertices that are equivalent to a given $v \in V$, then

$$l\,([v]) = [l(v)]$$

$$r\,([v]) = [r(v)]$$

Clearly, this definition is consistent, for if $u \equiv v$ and $v \in V-F$, then by the above definition: $u \notin F$, therefore $l(u) \equiv l(v)$ and $\tilde{l}\,([u]) = = [l(u)] = [l(v)] = \tilde{l}\,([v])$.

We could notice that the new structure induced by V/R, $\tilde{l}$, $\tilde{r}$ is no longer a forest, for instance the forest

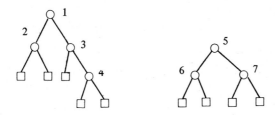

is trasformed in

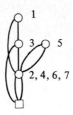

Definition 3. Given a compatible forest (V, F, l, r), we define the corresponding *linking,* as the graph with vertex set V/R and edge set:

$$\left\{ ([v], \tilde{l}([v])) \mid [v] \in (V{-}F)/R \right\} \cup \left\{ ([v], \tilde{r}([v])) \mid [v] \in (V{-}F)/R \right\}$$

Definition 4. Given a linked forest, we can define the following relation:

$$[u] < [v] \iff \exists \ u' \equiv u \text{ with } u' < v$$

It is easy to see that this is a *strict order* relation. This relation is characterized by the following property:

$$[v] < [v'] \Leftrightarrow \exists \, n > 0 \, \exists \, [v_0], \ldots, [v_n] \quad \left| \begin{array}{l} [v] = [v_0] \\[4pt] [v'] = [v_n] \\[6pt] \forall \ i < n \ [v_i] = \\ = \tilde{l}([v_{i+1}]) \text{ or } [v_i] = \\ = \tilde{r}([v_{i+1}]). \end{array} \right.$$

A consequence of this last property is the non existence of loops in the linked forest.

Definition 5. A *compatible linked forest (bec)* is an acyclic digraph that has the following properties:

1) There exists a unique vertex, called the *leaf,* which has no sons

(i.e. successors).

2) For any internal vertex (i.e. different of the leaf), its sons form an ordered set of cardinality 2. (We shall denote the elements of the set by left son and right son).

3) There are no two vertices having the same right son and left son.

Formally, we represent a *bec* by a quadruple $(V, l, r, \square)$ where V is the set of vertices, $\square \in V$ is the leaf, and l and r are two mappings:

$$l, r : \ V - \{\square\} \to V$$

which carry a vertex to its left and right sons, respectively.

Definition 6. Given a *bec*, we shall denote by *heads* those internal vertices which have no ancestors. Formally, the set of heads is given by:

$$C = V - (l(V) \cup r(V) \cup \{\square\})$$

2. Linking Algorithm*

We propose an algorithm for linking (in the sense of definition 3) any compatible forest. Let (V, F, l, r) be a given forest. We could make the linking procedure more effective if we assume that the set V of vertices in the forest has been previously classified by heights, where the height of a vertex is the length of the longest path from it to any leaf. Let H denote the height of the forest, which is defined as the maximum among the heights of all its vertices. Let $V(j)$ be the subset of vertices with height j. The following equations hold:

$$F = V(0)$$

$$V = \bigcup_{j=0}^{H} V(j)$$

* A preliminary version of this algorithm was presented in the 3rd Czechoslovak Symposium on Graph Theory. Praha, 1982.

The algorithm works by enumerating the class of equivalent vertices, assigning a 0 to the set of leaves, and increasing by on each new class. For each class i, it computes the value $\tilde{l}(i)$ and $\tilde{r}(i)$ of the mappings, in definition 2.

Linking algorithm

Input: H, $V(0) = F$, $V(1), \ldots, V(H)$, l, r.

Output: i, $\tilde{l}$, $\tilde{r}$ where $\tilde{l}$, $\tilde{r}$ are mappings from $\left\{ 1, \ldots, i \right\}$ into $\left\{ 0, 1, \ldots, i \right\}$ and NUM is a mapping from V into $\left\{ 0, \ldots, i \right\}$

begin

   for each vertex $u \in F$ *do* NUM $(u) \leftarrow 0$;

   $i \leftarrow 0$;

   for $j \leftarrow 1$ *until* H *do*

      $W \leftarrow V(j)$;

      $U \leftarrow \emptyset$;

      while $W \neq \emptyset$ *do*

            choose randomly a vertex v from W;

            delete v from W;

            if $u \in U$ s.t. NUM $(l(u))$ = NUM $(l(v))$ and

                           NUM $(r(u))$ = NUM $(r(v))$

            then NUM $(v) \leftarrow$ NUM (u)

            else $i \leftarrow i + 1$;

                  NUM $(v) \leftarrow i$;

                  $\tilde{l}(i) \leftarrow$ NUM $(l(v))$;

                  $\tilde{r}(i) \leftarrow$ NUM $(r(v))$;

                  add v to U

      endwhile

   endfor

end

3. Characterization of the *becs*

Proposition 1. A graph is a *bec* if and only if it is the linking of some compatible forest.

Proof

(i)   It is easy to prove from the previous definitions that the linking of every compatible forest gives a *bec*.

(ii) On the contrary, let us prove that every *bec* is generated by the linking of some compatible forest.

Let $G = (V, l, r, \square)$ be a *bec*. Let C be set of heads in G.

If $V = \{\square\}$ the result is trivial.

If $V \neq \{\square\}$ then $C \neq \emptyset$. Let $C = \{c_1, \ldots, c_k\}$

For each $c_i \in C$ we build a tree whose vertices are the set of all paths in G that go from c_i to some vertex in V. We shall notice that the paths differ not only in the vertices they contain but also in the edges. As a vertex could be at the same time the right and left son of other vertex, we shall use the following notation that differenciates the two cases. We shall use the sign "/" to separated two consecutive vertices of a path such that the second one is a left son of the first, and use "\" if the second vertex is a right son of the first vertex.

The Fig. 1 shows the construction of a *bec* from a forest consisting of two trees, and the construction to recover the original forest from the resulting *bec*.

The formal construction of the compatible forest corresponding to a given *bec* is easyly obtained by development of the ideas allready mentioned. On the other hand, it does not offer any difficulty to prove that the starting *bec* could be obtained from the forest which results from the linking.

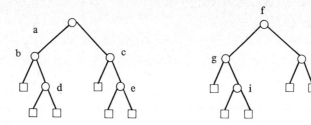

A forest:

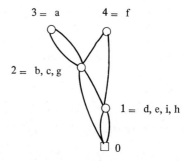

the *bec* corresponding
to the linkage of the
previous forest

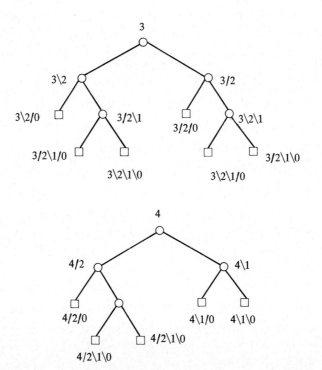

recovered forest from the previous *bec*

Fig. 1

4. Generation of *becs*

We are going to define a recursive equation and to prove later that it generates all *becs*.

Let us consider the following recursive equation defining a class X of diagraphs.

$$X = \square \cup \underset{X}{\left(\,\right)} \cup \underset{X}{\left(\,\right)} \cup \underset{X}{\left(\,\right)} \cup \underset{X}{\left(\,\right)} \cup \underset{X}{\left(\,\right)} \qquad [1]$$

In this equation each one of the six signs represents a generation rule, as it is done often in combinatorics (see for example [1] and [2]).

These differents rules are alternatives but they are not disjoints, therefore a given graph could have different generations with the same equation.

Let us look at the meaning of each rule.

Rule 1. The graph formed only by the leaf $\square$, belongs to the generated set.

Rule 2. The sign $\underset{X}{\left(\,\right)}$ represents the operation of adding a new vertex to the graph X, connecting it by the left and the right with any two vertices of X not heads (but they can be the leaf) in such a way that the new vertex would not be equivalent to any other vertex in X. (Recall that v is equivalent to u iff $l(u) = l(v)$ and $r(u) = r(v)$.)
For instance, the graph

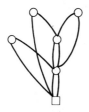

is obtained from

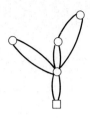

by application of rule 2.

Rule 3. The sign $\begin{smallmatrix} \bigcirc \\ X \end{smallmatrix}$ represents the operation of adding a new ver-

tex to the graph X, connecting it by left and right with some head in X (the same one for both connections).

For instance, the graph

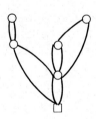

is obtained from

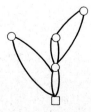

by application of rule 3.

Rule 4. The sign represents the operation of adding a vertex to

the graph X, connecting it by the left to a head of X and by the right
with a vertex of X different of a head.

For instance, the graph

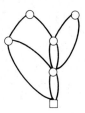

is obtained from

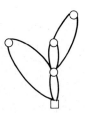

by application of rule 4.

Rule 5. The sign has an analogous meaning to rule 4 interchan-

ging left and right.

Rule 6. The sign represents the operation of adding a vertex to

the graph X, connecting it by the left and by the right to two differents heads in X.

An inductive argument on the total number of vertices shows that the graphs obtained by recursive generation using this set of rules fulfill all the requirements to be *becs*, and also that on the contrary, every *bec* can be generated by successive applications of these rules. Therefore we can state the following result

Proposition 2. Equation [1] generates exactly the set of *becs*.

5. Computation of the number of *becs*

Let us consider the following generating function

$$X(y, z) = \Sigma a_{k,n} y^k z^n$$

Where $a_{k,n}$ denotes the number of *becs* with k heads and n interior vertices.

To obtain an expression for this function we shall use formula [1], in the following way:

Rule 1. This rule introduces the *bec* with 0 heads and 0 interior vertices. From here we obtain $a_{0,0} = 1$, which is equivalent to

$$X(0, 0) = 1$$

Rule 2. This operation adds a new head and a new internal vertex to the *bec*. Therefore from each of the $a_{k-1, n-1}$ *becs* with $k-1$ heads and $n-1$ interior vertices, we obtain $(n-k+1)^2 - n + 1$ *becs* with k heads and n interior vertices, because there are $n-k+1$ vertices not heads, to which we can connect the new vertex, and it is necessary to eliminate the $n-1$ possibilities that would make it equivalent to some of the existent vertices.

Thus, this operation contributes with the term

$$[(n-k+1)^2 - n + 1] a_{k-1, n-1} y^k z^n$$

which is the term of order $y^k z^n$ in the expression:

$$yz^3 \frac{\partial^2 X}{\partial z^2} + y\, z\, \frac{\partial^2 X}{\partial y^2} - 2y^2 z^2 \frac{\partial^2 X}{\partial y \partial z} +$$

$$+ 2yz^2 \frac{\partial X}{\partial z} - y^2 z \frac{\partial X}{\partial y} + yz$$

Rule 3. This operation adds a new interior vertex and leaves without changing the total number of heads in the original *bec*. Thus, from each one of the $a_{k,n-1}$ *becs* with k heads and $n-1$ interior vertices we could obtain k *becs* with k heads and n interior vertices, one for each existent head.

Therefore, this operation contributes with the term:

$$k.a_{k,n-1} y^k z^n$$

that is the term of order $y^k z^n$ in the expression:

$$yz\, \frac{\partial X}{\partial y}$$

Rules 4 and 5. Each one of these operations adds a new interior vertex and leaves invariant the total number of heads in the original *bec*. Thus, from each one of the $a_{k,n-1}$ *becs* with k heads and $n-1$ interior vertices we obtain $k(n-k)$ *becs* with k heads and n interior vertices.

These operations add the term:

$$2\,.\,k.\,(n-k)\,.\,a_{k,n-1}\,.\,y^k\,.\,z^n$$

which is the term of order $y^z z^n$ in the expression:

$$- 2y^2 z \frac{\partial^2 X}{\partial y \partial z} + 2yz^2 \frac{\partial^2}{\partial y} \frac{X}{\partial z}$$

Rules 6. This operation adds a new vertex and decreases by one the total number of heads in the *bec*. From each one of the $a_{k+1,n-1}$ *becs* of $k+1$ and $n-1$ interior vertices, we obtain $(k+1)$ k *becs* with k heads and n internal vertices, which correspond to the different ways to connecting the new head to 2 different heads.

Therefore, this operation contributes with the term:

$$(k+1) . k . a_{k+1,n-1} . y^k z^n$$

which is the term of order $y^k z^n$ in the expression:

$$yz \frac{\partial^2 X}{\partial y^2}$$

Finally, it is necessary to recall that this generation rules are not disjoint. Therefore, to count the number of *becs* with k heads and n interval vertices, if we start from the addition of the above expressions, some *becs* will be counted more than once. More precisely, each *bec* will be counted as many times as the number of different ways of generating it by use of formula [1]. But, all rules except the first one, add a new head to an existing *bec*.

Moreover, if we only consider the operations wich introduce a new head, it could be observed that a *bec* of k heads $(k > 0)$ and n vertices can be obtained in k different ways from the original *bec* of $n-1$ vertices. Therefore, the addition of the contributed terms due to rules 2 to 6 must be equal to:

$$k . a_{k,n} y^k z^n$$

which is the term of order $y^k z^n$ in the expression:

$$y \cdot \frac{\partial X}{\partial y}$$

The equivalence of this expression with the addition of the ones which come from rules 2 to 6 stablishes the following result.

Theorem. The generating function $X(y, z) = \sum a_{k,n} y^k z^n$ where $a_{k,n}$ is the number of *becs* with k heads and n vertices, is defined by the following equation

$$z(y-1)^2 \frac{\partial^2 X}{\partial y^2} - 2z^2(y-1) \frac{\partial^2 X}{\partial y \partial z} + z^3 \frac{\partial^2 X}{\partial z^2} -$$

$$- (ys-z+1) \frac{\partial X}{\partial y} + 2 z^2 \frac{\partial X}{\partial z} + zX = 0$$

with the initial condition: $X(0, 0) = 1$

The integration of this equation is an open problem. In table 1, are given the values of the first coefficients $a_{k,n}$ in the expansion of X.

References

[1]   Comtet, L.: *"Analyse combinatoire"*, 2 vol. P.U.F. Paris (1970).

[2]   Flajolet, P.; J. M. Steyaert: *"A Complexity Calculus for Classes of Recursive Search Programs over Tree Structures"* Proc. of IEEE 22nd. FOCS Sym., Nashville TN(1981) pp. 386-393.

[3]   Reingold, E. M.; J. Nievergelt; N. Deo: *"Combinatorial Algorithms; Theory and Practice"*. Prentice Hall, N. J. (1977).

k ╲ n	0	1	2	3	4	5	6	7	8	9
0	1	0	0	0	0	0	0	0	0	0
1		1	3	15	111	1119	14487	230943	4395855	97608831
2				3	60	1089	21306	465855	11439648	314210145
3					1	105	4728	177910	6510495	244633911
4							105	12555	913905	55741620
5								63	23040	3314763
6									21	30867
7										3

Table 1

Annals of Discrete Mathematics 25 (1985) 125-144
© Elsevier Science Publishers B.V. (North-Holland)

RIGID EXTENSIONS OF GRAPH MAPS

I. S. FILOTTI

L.R.I. – Université de Paris – Sud, Orsay (France)

1. Introduction

A graph is γ-rigid if it admits a unique embedding in the closed orientable surface of genus γ.

A first remarkable discovery concerning this notion is due to Whitney [7] who proved that 3-connected planar graphs are 0-rigid. His original proof is rather involved. A considerably shorter one was folklore and appeared for the first time, to our knowledge, in [5], (problem 6.69). In [4] the notion of rigidity of maps was studied and generalized in connection with algorithms for determining the isomorphism of graphs of given genus.

One sees easily that for $\gamma > 0$ the theorem of Whitney no longer holds. Consider, however, the *extension* problem which is the following: given a map H and a supergraph G of its skeleton, to determine whether H has an extension of the same genus as H to G. If this extension problem has a unique solution we shall say that it is *rigid*. The generalization of Whitney's theorem given in [4] claims that if G is 3-connected and if H is simplicial and a *frame* then the extension problem (H, G) is rigid. Frames will be defined in the body of this paper. In [4] a proof of this generalization was sketched. It paralleled the original one of Whitney.

On the other hand, Demoucron, Malgrange and Pertuiset [1] had

This work was supported by the National Science Foundation under Grant MCS–7924401

proposed a planarity algorithm that seems to be the simplest and most natural one to date. In this paper a generalization of this algorithm, is presented which can be used to give simpler proofs of both Whitney's theorem and of its generalization.

Our terminology is mostly standard. Graphs have no self-loops and no multiple edges. Vertices are sometimes called points. If G is a subgraph of H we write $G \subset H$ and we also call H a supergraph of G. $H - G$ will denote the subgraph of H induced by the edges of H that are not in G. G determines subgraphs called *bridges*. These are equivalence classes of the following equivalence relation on the edges of H: $e \sim è$ if $e = e'$ or if there is a path connecting e to e' disjoint from G (see e.g. [5]. The points on G of a bridge are its *attachments*.

For notions of topological graph theory ˙one can consult [6]. An embedding of a graph in an orientable surface is specified combinatorially by cyclical orderings of the edges incident to each vertex. *Embedding* and *map* are synonims. The two longitudinal orientations of an edge will be called its *sides*. We usually denote a graph by superscript[1] as in G^1. Maps are indicated by superscript[2] as in G^2. The graph underlying a map G^2 is called its *skeleton* and is denoted G^1 or sk (G^2).

A face is *simplicial* if every vertex appears only once on it. A simplicial map is one with only simplicial faces. α_i (K) is the number of simplices of dimension i of complex K (i = 0, 1, 2). The Euler characteristic of a map G^2 is $\chi(G^2) = \alpha_0(G^2) - \alpha_1(G^2) + \alpha_2(G^2)$. It is always even, is /2 for planar maps and 0 for toroidal ones. $\chi(G^2) = 2 - 2\gamma(G^2)$. $\gamma(G^2)$ is the *genus* of G^2.

Every map has an *inverse* obtained by taking the inverse cyclical ordering at each vertex. We shall identify a map with its inverse.

A map G^2 indices a map H^2 on a subgraph H^1 of G^1 by restricting the cyclical order at each vertex. H^2 will be called a restriction of G^2 and G^2 an extension of H^2. We shall often write $G^2 \uparrow H^1$ to denote the restriction H^2. In general $\gamma(H^2)$ is no bigger than $\gamma(G^2)$, but if $\gamma(H^2) = \gamma(G^2)$ the extension is said to be *conservative* and we write $H^2 \subset G^2$.

If $H^2 \subset G^2$, the bridges of G^1 w.r.t. H^1 are subcomplexes of the faces of H^2. We shall say that they are embedded or lie in those faces. Combinatorially, a chain can be said to be embedded in a face if the ends are inserted between two successive edges of the face.

A map is a *frame* if no pair of faces contains an essential, i.e. non-contractible cycle.

2. Conservative extensions

We shall need the following

Lemma. Let H^2 be a map and G^2 a conservative extension. Then every bridge lies within a face of H^2.

Proof. Assume not. Then there exists a chain connecting two faces and whose ends are inserted between successive edges in H^2 of these faces. Let K^2 be the map obtained from H^2 by so modifying the cyclical ordering of the edges. We may assume that the chain consists of a single edge.

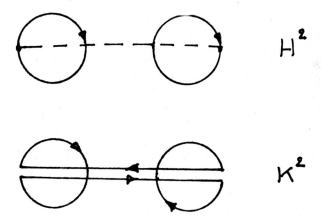

Fig. 1

A simple calculation shows that in K^2 the number of points stays the same as in H^2, while the faces decrease and the edges increase. Hence the genus of K^2 increases by 1 as does that of G^2.

3. Frames

In a frame two faces that meet contain no essential cycle and hence are contained in a disk.

Lemma 1. Let F_1 and F_2 be two faces that touch in a frame H^2. Let K^2 be the map induced on the subgraph of H^1 consisting of faces F_1 and F_2. In K^2 any face that meets both F^1 and F^2 has no more than two points in common with both.

Proof. Let F_3 be a face of K^2 that has three vertices in common with F_1 and F_2. K^2 is planar. In a planar embedding the cyclic orientation of three vertices common to two faces are opposite. Hence the three vertices cannot belong to more than two faces.

Corollary. In a frame three faces cannot share three points.

The lemma and its corollary give the general shape of a pair of faces in a frame.

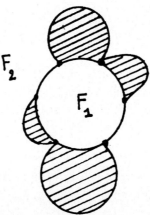

Fig. 2

Two vertices of F_1 and F_2 that also lie on one of the shaded faces are connected by two disjoint paths, one on F_1 and the other on F_2.

We now claim that one can remove one of these chains and that the resulting map is still a frame. Assume otherwise. Let C be the chain removed and let us assume for definiteness that C borders F_2. In the new map F_2 is slightly extended to F_2'. An essential cycle that would exist now and traverse a pair of faces would have to traverse F_3. It is easy to see then that the original map would also contain an essential cycle, namely the one obtained by replacing the portion that crosses F_3 by the part of F_3 that connects the two intersection points of the path with F_3. We have thus proved

Lemma 2. A frame F^2 contains a subframe K^2 with the property that the intersection of two faces has no more than two connected components.

A frame with the property of the lemma shall be called *reduced.*

4. Hindering

Let H^2 be a map and $G^1 = H^1 \cup B$ be an extension by a bridge attached to a face F of H^2. We shall say that B is *embeddable* in F if there exists an extension G^2 of H^2 to G^1 in which F no longer is a face. Likewise, two bridges attached to F will be said to be *simultaneously embeddable* in F if there exists an extension such that all the faces but F are faces of the extension as well.

A sufficient condition for simultaneous embeddability is given by the following creiterion of Demoucron, Malgrange and Pertuiset [1].

Say that to chains B1 with extremities x_1 and y_1 and B_2 with extremities x_2 and y_2 *overlap* if their extremities appear in the order x_1, x_2, y_1, y_2 on the same face. F.

Lemma. Two bridges attached to face F each individually embeddable in it are not embeddable simultaneously iff either:

(i)   they posses three common attachments, or

(ii)   they posses overlapping chains.

Two bridges satisfying the condition of the lemma will be said to *hinder* each other. We shall then write $B_1 \mid_F$ raise B_2

The simple criterion provided by the lemma is purely combinatorial and does not require the inspection of the whole bridge but just of its attachments. It is essential for our algorithm.

5. The compatibility graph

Let H^2 be a map and let G^1 be a supergraph of H^1. To this extension problem we shall associate a labeled graph denoted by Comp (H^2, G^1) called the *compatibility graph* associated to the extension problem (H^2, G^1). In Comp (H^2, G^1) there will be a vertex for every bridge of $G^1 - H^1$, labeled with the name of the bridge and with the names of all the faces containing all the attachments of the bridge. A pair of vertices will be connected by an edge if the corresponding bridges hinder each other. The edge will be labeled with the names of all the faces with respect to which the extremities hinder each other.

A *satisfying assignment* is a map assigning a face to each vertex of the compatibility graph, i.e. to each bridge, such that:

(i)   the bridge has all its attachments on the face, i.e. the face is among the labels decorating the corresponding vertex,

(ii)   adjacent vertices may bear the same label only if it does not appear on the edge joining them. This expresses the fact that we do not want to embed in the same face bridges that hinder each other.

We shall say that a bridge B of $G^1 - H^1$ is *forced* if it has all its attachments on a single face of H^2. By extension, we shall say that every subgraph of a forced bridge is forced (to that face). A bridge will be said to have as many choices as there are faces that contain all its attachments, i.e. as there are labels decorating it in the compatibility graph.

Clearly, if the extension problem has a solution there also exists a satifying assignment for the compatibility graph, namely that which assigns to every bridge the face in which it has been embedded in the extension.

The converse does not hold. It does not suffice to have a satisfying assignment for the compatibility graph for the extension to exist. One also has to check that each bridge can be embedded in the face that has been assigned to it. Checking whether the compatibility graph can be satisfied is however a mandatory step in our algorithm.

If the bridges always possess an embedding in the face no verification is needed beyond the existence of the satisfying assignment. This is the case for "stars", i.e. trees in which all vertices but at most one have degree one (such vertices are sometimes called leaves).

Call two bridges *equivalent* if they share all their attachments. A set of equivalent bridges with just two attachments will be called a "mellon".

Lemma 1. Let $K^1 \supset H^1$ be an extension by stars of H^1 and let α be a satisfying assignment for $\text{Comp}(H^2, K^1)$. Then

(i) there exists an extension K^2 of H^2 agreeing with α, i.e. assigning to each bridge the same face as α.

(ii) if H^2 is simplicial and there are no mellons and if further G^2 is 2-connected, then the extension is unique.

Proof.

(i) Stars have an embedding in any face containing all their attachments. The existence of α guarantees that no two will hinder each other.

(ii) Any star has a unique embedding in a simplicial face. We therefore only have to check that given two bridges B_1 and B_2 there is only one joint embedding of B_1 and B_2 in a face to which α simultanously assigns them. We claim that this is indeed the case if there are no mellons.

There are two cases to consider. Either one of the bridges, say B_1, has three attachments on H^1 or both have two attachments (the case of just one attachment is eliminated since the graph is 2-connected).

If B_1, say, has three (or more) attachments to H^1 it splits F into three subfaces and the attachments of B^2 can lie on at most one of them.

If B_1 and B_2 both have two attachments there must be at least three different ones, since there are no mellons and since it is obvious that they admit of only one embedding in F.

It will be convenient to identify the different embeddings of a mellon in a face. With this convention uniquenes is guaranteed in all cases.

The next lemma, the Confluence Lemma, shows that once a satisfying assignment has been chosen one can proceed by selecting a chain in one of the bridges and embed it in the face assigned to it. The choice of bridge does not matter, as does not the choice of chain in the bridge.

Let $H^2 \subset K^2$ be an extension and let α be a satisfying assignment of $\text{Comp}(H^2, K^1)$. If K^2 embeds the bridges of $K^1 - H^1$ in the faces assigned to them by α we shall say that K^2 *agrees* or is compatible with it.

Lemma 2 (Confluence Lemma). Let C_1 and C_2 be chains of G^1 w.r.t. H^1, spanning H^1, and let $\alpha \in \text{Comp}(H^2, G^1)$. Let $K_i^1 = H^1 \cup C_i$ $(i = 1, 2)$ and let K_i^2 be extensions of H^2 compatible with α. Then K_1^2 has an extension to a map G^2 of G^1 if and only if K_2^2 does.

Proof. Assume K_1^2 has an extension compatible with α to a map G^2 of G^1. Then G^2 is compatible with α and hence $G^2 \uparrow K_2^1 = K_2^2$ since by Lemma 1 there is a unique map of K_2^1 compatible with α. Therefore K_2^1 can be extended to G^2.

6. Extension of simplicial maps

We now consider the extension of a simplicial map. This case is interesting because a chain can be embedded in a face in only one way. If
the embedding is not simplicial a face may contain both sides of an
edge or more than one instance of a vertex. Then a chain can have
more than one embedding depending on the instances it connects. For
example, in figure 3, a and b represent two different embeddings of
the same chain.

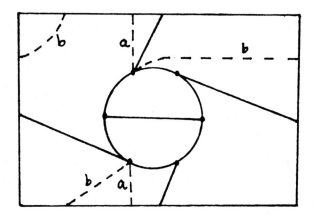

Solid lines represent an embedding of $K_{3,3}$ in the torus. The torus is represented by the outside rectangle with opposite edges identified as indicated by the arrows.

Fig. 3

The following algorithm solves the extension problem (H^2, G^1) in
the case of a 2-connected graph G^1 and of a simplicial map H^2.

Main extension algorithm for simplicial maps

INPUT : Simplicial map H^2, 2-connected graph $G^1 \supset H^1$.

OUTPUT : Extension G^2 of H^2 with skeleton G^1, if one exists.

1. If $H^1 = G^1$, stop and output H^2.
2. Construct the compatibility graph associated to (H^2, G^1).
3. Find all satisfying assignments for Comp (H^2, G^1).
4. For each satisfying assignment do:

    4.1.   Pick an arbitrary bridge and the face assigned to it.

    4.2.   Pick a chain with distinct end points spanning the bridge. Embed it in the chosen face. Let K^2 be the resulting map.

    4.3.   Recursively call the algorithm with (K^2, G^1).

One can also give a parallel version of this algorithm where one selects simultaneously a chain in each bridge and embeds it in the face assigned to it.

The correctness of this algorithm is easy. If there exists an extension then there is also a satisfying assignment. We check them all. The Confluence Lemma now shows that the choice of chain does not matter. We call the algorithm recursively. For this to be possible the new map obtained after embedding the chain must also be simplicial. It, is because extending a map by a chain does not destroy simpliciality. Finally, we observe that in step 4.2 it is always possible to pick a chain with distinct extremities since G^1 is 2-connected.

The running time is determined by steps 3 and 4.3. Since at each recursive call the number of edges of $G^1 - H^1$ decreases by at least 1, the number of recursive calls is bounded above by the number of edges in $G^1 - H^1$. Let $n = \alpha^1(G^1 - H^1)$ and let $t(n)$ be the maximum running time in this case. Likewise, let $t_1(n)$ be the time required to find all satisfying assignments for $\text{Comp}(H^2, G^1)$ and let $t_2(n)$ be the maximum number of satisfying assignments. We can then write:

$$t(n) = t_1(n) + t_2(n) \cdot t(n-1)$$

We have showed elsewhere that the problem of determining the satisfying assignments is NP—complete being easily seen to be reducible to the problem of finding all satisfying assignments of a propositional formula in conjunctive normal form. Thus both $t_1(n)$ and $t_2(n)$ seem to be of an exponential nature. The main algorithm is therefore not a practical one.

7. The DMP algorithm

In special cases one can obtain much better running times than for the main algorithm. The original Demoucron, Malgrange and Pertuiset algorithm applied to the planar case. It can be generalized to extensions of frames. This is the version we shall present in this section.

Throughout this section H^2 will be a frame and G^1 a 2-connected supergraph of its skeleton H^1.

Theorem 1 (Indifference Theorem).

(i)   A bridge B of $G^1 - H^1$ with more than one face containing all its attachments is embeddable in one of them iff it is in the other.

(ii)   Two adjacent vertices of $Comp(H^2, G^1)$ have no more than two common labels.

(iii) Let B_i $(i = 1,2)$ be embeddable in faces F and F_i, where $F \neq F_i$. Assume that $B_1 \mid_F B_2$. Then $F_1 = F' = F_2$ and $B_1 \mid_{F'} B_2$.

Proof.

(i)   Any two faces in which B is embeddable contain no essential cycle, hence can be contained in a disk. The attachments of B to one of them must appear in the apposite order on the other. The claim now follows since the inverse of a map changes the cyclical ordering of an embedding in its inverse.

(ii)   This is just a reformulation of the Corollary to Lemma 1 of section 3.

(iii) There are three cases to consider:

Case 1. B_1 and B_2 have three common attachments. If F, F_1 and F_2 were mutually distinct there would be three distinct faces meeting at three points, contradicting (ii). Hence $F_1 = F' = F_2$ and $B_1 \mid_F' B_2$.

Case 2. B_1 and B_2 contain overlapping chains with ends a_i, b_i (i=1,2). We may assume by Lemma 2 of section 2 that the intersection of any two faces has at most one connected component.

Assume now that B_i is not embeddable in F_j ($i \neq j$; i, j = 1, 2).

Since a_1, $b_1 \in F_1$, one of a_2 and b_2, say b_2, lies on F_1. Likewise, either a_1 or b_1, say b_1 must lie on F_2. Therefore, a_2 and b_1 lie on all three faces F, F_1, and F_2. But the three faces cannot have three vertices in common.

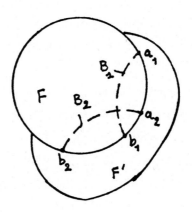

Fig. 4

Thus B_1 and B_2 are both embeddable in both F_1 and F_2. But then F, F_1, and F_2 have more than two points in common. This is only possible if $F_1 = F_2$.

The theorem is not valid if H^2 is not a frame. For example, assume that H^2 is the simplicial map of $K_{3,3}$ shown in solid lines in figure 5.

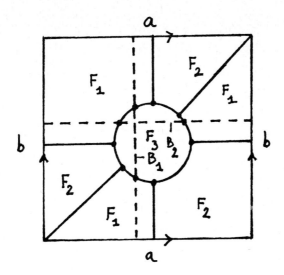

Fig. 5

Here B_1 and B_2 hinder each other with respect to F_3 but not with respect to F_1, although they are embeddable in either.

We now give the structure of the compatibility graph.

Lemma 2.

(i)   In a connected component of the compatibility graph with more than one vertex every vertex and edge have at most two labels.

(ii) In a connected component with a single vertex, the vertex may have an arbitrary number of labels.

(iii) If all vertices of a connected component of the compatibility

graph have two labels, then all vertices and edges of the component bear the same pair of labels.

Proof. (i) and (iii) follow trivially from the Indifference Theorem.

Case (ii) can occur in the case of bridges with one or with two
attachments. The case with one attachment is relatively trivial since
we usually consider 2-connected graphs. We shall call the "mellon"
case the case of a bridge with two attachments attached to many
faces.

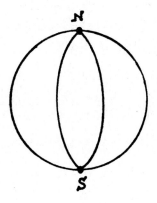

Fig. 6

Corollary 1. Assume that all bridges have two choices. If H^2 admits
an extension G^2 it also admits the extension in which every bridge is
embedded in the other face than the one it is embedded in G^2.

Proof. The compatibility graph is bipartite and hence the assignment
that assigns to every bridge the other face than in G^2 is also satisfying.
Further, by the lemma, if a bridge is embeddable in a face it also is
embeddable in the other face that has all its attachments.

Corollary 2. Assume that in $Comp(H^2, G^1)$ all bridges have two
choices. Let B be a bridge attached to two faces F_1 and F_2 and let C

be a chain spanning B and with distinct extremities on H^2. Let K_i^2 ($i = 1, 2$) be the map obtained by embedding C in F_i. Then K_1^2 has an extension to G^1 iff K_2^2 does.

Proof. Follows from the previous Corollary and the Confluence Lemma.

Lemma 3. Assume that $Comp(H^2, G^1)$ has a forced bridge B. Let C be a chain spanning B. Let K^2 be the extension of H^2 to $K^1 = H^1 \cup C$ that embeds C in the forced face. Let G^2 be an extension of H^2 to G^1. Then $G^2 \uparrow (H^1 \cup C) = K^2$.

Proof. Let α be the satisfying assignment for $Comp(H^2, G^1)$ induced by G^2. Since B is a forced bridge, α must agree with that forced choice on B.

These considerations lead us to the following algorithm.

The DMP algorithm

INPUT : Frame H^2, 2-connected graph $G^1 \supset H^1$ raised 1.

OUTPUT : Extension G^2 of H^2, if one exists.

1. If $H^1 = G^1$, halt and output H^2.

2. Pick a bridge and a face as follows:

   - if there exists a forced choice (i.e. a bridge with just a face containing all its attachments), pick such a bridge and assign it that face.
   - if there exist only bridges with two choices, pick any bridge and assign it any of the faces that contain all its attachments.

3. Pick a chain with distinct extremities spanning the bridge and embed it in the chosen face. Let K^2 be the new map.

4. Recursively call the algorithm with (K^2, G^1).

A tower

$$H^2 = H_0{}^2 \subset H_1{}^2 \subset \ldots \subset H_h{}^2 = G^2$$

of extensions will be called a DMP sequence if H^2_{i+2} is an extension by one chain of $H_i{}^2$ ($i = 0, 1, \ldots, h\text{-}1$) satisfying the conditions of the DMP algorithm.

Theorem 2 (Correctness of the DMP algorithm).
If the extension problem (H^2, G^1) has a solution G^2, then there exists a DMP sequence with G^2 as its last term.

Proof. We shall show how to construct the term H^2_{i+1} from H_i. Just apply the DMP algorithm. As long as there are no choices to make, there is nothing to specify. The only case where there are choices is that in which the compatibility graph has no forced choices. In that case select any bridge and assign it the face assigned to it by G^2.

We now claim that $G^2 \uparrow H_i{}^1 = H_i{}^2$ and that G^2 is an extension of $H_i{}^2$. The proof is by induction on i and is completely trivial.

This algorithm differs from the main one presented earlier in that no satisfying assignment for the compatibility graph is ever calculated. Instead we directly pick a bridge and a face bypassing step 3.

This results in a considerably shorter running time. Iedeed, if n is the number of edges of $G^1 - H^1$ then there are at most n recursive calls. A sloppy implementation will require n steps for each call resulting in an $0(n^2)$ algorithm. However, an $0(n)$ implementation is possible. We shall not give this implementation here and be content with noticing that the algorithm certainly runs in polynomial time.

8. The extension problem for 3-connected graphs

In the case of 3-connected graphs the extension problem is in a sense simplified, since there is at most one extension of a frame. This, of course, is the contents of the rigidity theorem.

Let H^2 be a frame (that is not a cycle) and let G^1 be a 3-connected supergraph of its skeleton. We may assume without loss that H^2 is reduced and that it admits (at least one) conservative extension G^2 to G^1.

Lemma 1. Let the extension problem (H^2, G^1) admit a DMP sequence all terms of which are forced. Then its last term is independent of the sequence.

Proof. Let $H^2 = H_0{}^2 \subset H_1{}^2 \subset \ldots \subset H_h{}^2 = G^2$ be a DMP sequence terminating in G^2. Assume that H^2 admits an extension $G_1^2 \neq G^2$. We shall show by induction on i that $H_i{}^2$ also admits an extension to $G_1{}^2$. This will prove that $G_1{}^2 = G^2$. For i = 0, the claim is obvious by hypothesis. For i > 0, H_{i+1}^2 is a forced extension of $H_i{}^2$, and must be also extensible to both $G_1{}^2$ and to G^2.

To prove the rigidity theorem it now suffices to show that there exists a DMP sequence of forced choices leading to G^2.

Lemma 2. For any conservative extension G^2 of H^2 there exists a DMP sequence of forced choices whose last term is G^2 and all of whose terms are reduced frames.

Proof. We begin by showing that there exists a DMP sequence all terms of which are forced. It suffices to show it for H^2.

If $Comp(H^2, G^1)$ has no forced choices, then, by the structure theorem there are two kinds of components of $Comp(H^2, G^1)$:

(i) Components all vertices and edges of which have precisely two labels F_1 and F_2 and

(ii) Components with just one vertex (and perhaps more than two labels).

Case (i). Let F_1 and F_2 be the two faces. Since $F_1 \cap F_2$ has a single connected component which must be a chain, there are two (perhaps coinciding) extreme attachments of a bridge to $F_1 \cap F_2$. The bridge or $F_1 \cap F_2$ must have at least one internal vertex. Then the graph can be disconnected by removing just two vertices contradicting the 3-connectivity of G^1.

Case (ii). This is the "mellon" case. Let N and S be the two vertices of $F_1 \cap F_2$. A bridge B connected to just N and S must be an edge since otherwise the graph can be disconnected by removing N and S and this would again violate the 3-connectivity of G^1.

Since B is just an edge all the chains running from N to S in the mellon must contain internal vertices. The mellon cannot contain a single face since otherwise B would have all its attachments on a single face and hence there would be forced choices.

The mellon thus contains at least two faces and hence at least an edge with an internal vertex common to two faces of the mellon.

This internal vertex must be connected to some other vertex of the mellon and in fact must be an internal vertex of a chain running from N to S. But this chain would hinder B, against our hypothesis.

We have thus proved that there always exists a forced choice. Let B be a bridge with a forced choice F. Then there must exist two other faces F_1 and F_2 such that B has an attachment on each. Then choose a chain C spanning B and connecting those two attachments. This will guarantee that H_1^2 is a frame.

We can now state formally.

Theorem (Rigidity Theorem). Given a frame H^2 and a 3-connected supergraph G^1 of H^1, there exists at most one conservative extension G^2 of H^2.

Proof. By the preceding lemma. If the graph is not reduced, reduce it first and then apply the lemmas.

In the planar case it is necessary to assume that the frame is not a cycle, since otherwise there are trivially two choices (the inside and the outside of the face constituted by the cycle). Whitney's theorem then follows immediately.

ACKNOWLEDGEMENTS

I am very grateful to Howard Levene for his support.

References

[1] DEMOUCRON, G. Y. MALGRANGE AND R. PERTUISET, Graphes Planaires: Reconnaissance et Construction de Représentations Planaires Topologiques, Rev. Française Informatique et Recherche Opérationnelle 8 (1964), 33-47.

[2] I. S. Filotti and G. L. Miller, *"Efficient Determination of the Genus of a Graph (preliminary version)"*, Proceedings, Eleventh Annual ACM Symposium on the Theory of Computing, pp. 27-37, Assoc. for Computing Machinery, N. Y. 1979.

[3] I. S. Filotti, *"An Algorithm for Embedding Cubic Graphs in the Torus"*, J. of Computer and System Sciences, 20, (1980), 2, 255-276.

[4] I. S. Filotti and J. N. Mayer, *"A Polynomial-time Algorithm for Determining the Isomorphism of Graphs of fixed Genus (preliminary version)"*, Proceedings, Twelfth Annual ACM Symposium on the Theory of Computing, pp. 236-242, Assoc. for Computing Machinery, N. Y. 1980.

[5] LOVASZ, L. *"Combinatorial Problems and Exercises"*, North-Holland Publ. Co., Amsterdam, 1979.

[6] WHITE, A. T. *"Graphs, Groups and Surfaces"*, North-Holland/American Elsevier, New York, 1973.

[7] WHITNEY, H. Congruent Graphs and the Connectivity of Graphs, American J. of Math., 54 (1932), 150-168.

Annals of Discrete Mathematics 25 (1985) 145-188
© Elsevier Science Publishers B.V. (North-Holland)

ALGEBRAIC METHODS FOR TRIE STATISTICS

Philippe FLAJOLET

INRIA Rocquencourt 78153-Le Chesnay (France)

Mireille REGNIER

INRIA Rocquencourt 78153-Le Chesnay (France)

Dominique SOTTEAU

LRI Universite Paris-Sud 91405-Orsay (France)

Tries are a data structure commonly used to represent sets of binary data. They also constitute a convenient way of modelling a number of algorithms to factorise polynomials, to implement communication protocols or to access files on disk. We present here a systematic method for analysing, in the average case, trie parameters through generating functions and conclude with several applications.

1. Introduction

Digital Searching methods comprise a variety of techniques used for sorting or retrieving data by taking advantage of their binary representations. In many cases, these techniques constitute an attractive alternative to comparison-based methods that relie on the existence of an ordering on the universe of data to be processed.

The *trie* structure is probably the most well-known amongst digital structures. It is a tree representation of sets of digital (*e.g.* binary) sequences that has been introduced by de la Briandais and Fredkin [9]† and bears analogies to binary search trees [11, 15, 17, 20, 1, 19] Operations like insert, delete, query, union, intersection... can be performed efficiently on this representation of sets. Tries also appear as a structure underlying *Radix Exchange Sort* (a digital analogue of Quicksort, [11], p. 131).

† *The name* trie *was coined by Fredkin apparently from a combination of* tree *and re*trieval.

Tries have been proposed as an efficient way of maintaining *indexes* for externally stored files. When combined with hashing techniques (to ensure a uniform distribution of elements on which tries are built) they lead to *dynamic hashing* schemes like the Dynamic Hashing Method of [12] or Extendible Hashing [2].

Another use of tries is for *multi-dimensional* searching; the problem there is to retrieve records with several fields when only some of these fields are specified in a query. Under the form of *k-d-tries* (multi-dimensional tries) they constitute an elegant solution to the problem of *Partial Match* or *Secondary Key* retrieval. This variety of tries has been described by Rivest [18] who assigns their origin to Mc Creight. Used in conjunction with ideas taken from dynamic hashing techniques they lead to the so-called *Grid-File* algorithms [14] that have been proposed as a physical access method for files.

As a representation of binary sequences, tries are also of frequent occurrence in several applications. As an example, Huffman's algorithm may be viewed as a progressive construction of a trie. Situations where they appear to be a convenient model are for instance polynomial factorisation [8], communication protocols [3] or some simulation algorithms [6]. In many such cases, rather intricate parameters of binary sequences have simple formulations when expressed in terms of tries.

Our objective here is to describe a general set-up in which statistical analyses on tries can be conveniently performed. We show how to derive in a concise and synthetic way generating functions for average values of a large number of parameters of interest in the context of the analysis of algorithms. In this manner, we are able to present in a systematic manner (and sometimes extend) a number of analyses otherwise often obtained at some computational effort, and show that they can be reduced to a few simple paradigms.

Our methodology consists in first establishing a few basic and easy to prove lemmas; these lemmas, given in Sections 2,3, relate under various probabilistic models some structural definitions of trie parameters to functional equations of some sort over generating functions of average values for which general resolution methods are also available. We thus have an *algebra of parameters* on tries which is mapped on an *algebra of generating functions.* In this way, the process of analysis is reduced to finding expressions for parameters of interest as combinations of a few building blocks for which mapping lemmas are available, and obtaining generating function expressions (whence expressions for average values) becomes an almost mechanical process. Most notably, the recourse to recurrences on average values which constitutes the basic technique usually employed is completely eliminated. This permits to analyze in a simple way rather complex parameters of tries. The usefulness of this approach is demonstrated on several examples in Sections 4,5.

We do not address here the problem of the asymptotic evaluation of trie parameters (*cf* [11], p. 131), but occasionally mention some of the estimates to make clear the implications of the analyses. The key method there consists in using Mellin transform techniques and some systematisation of its use is also possible [7].

1.1. Trie Representations of Sets

Assume we want to represent data from a universe of *binary strings,* sometimes called *records* or *keys,* of a fixed length $s \geq 0$:

$$\mathbf{B}^{(s)} = \left\{ 0; 1 \right\}^s$$

A subset $\omega \in \mathbf{B}^{(s)}$ decomposes into two subsets $\omega/0$, $\omega/1$ of $\mathbf{B}^{(s-1)}$ defined by:

$$\omega/0 = \left\{ v \in \mathbf{B}^{(s-1)} \mid 0v \in \omega \right\}$$
$$\omega/1 = \left\{ v \in \mathbf{B}^{(s-1)} \mid 1v \in \omega \right\}.$$

The definition of a trie is based on this decomposition.

Definition: *To a subset $\omega \subset \mathbf{B}^{(s)}$, we associate a tree trie (ω) as follows:*

(1) *If* $|\omega| = 0$, *then trie* (ω) *is the empty tree.* †

(2) *If* $|\omega| = 1$, *then trie* (ω) *is a tree formed with a unique leaf labelled* ω.

(3) *If* $|\omega| \geq 2$, *then trie* (ω) *is obtained by appending a root to the recursively defined subtrees trie* ($\omega/0$) *and trie* ($\omega/1$).

As an example, the trie associated to:

$$\omega = \left\{ a, b, c, d, e, f \right\}$$

with

$a = 01011$, $b = 01101$, $c = 10110$, $d = 11000$, $e = 11011$, $f = 11110$

is displayed in Figure 1.

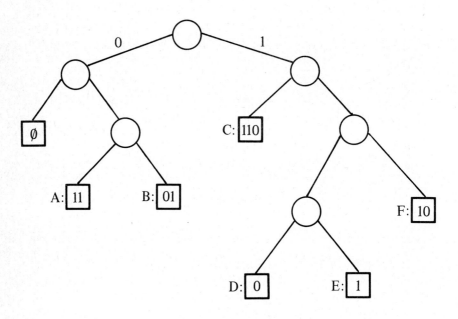

Fig. 1: A trie constructed on 6 binary sequences of length 5.

†*We let* $|\omega|$ *denote the number of elements (cardinality) of* ω

The way of constructing the trie is clear form the definition: we recursively partition the set to be represented according to bits of highest weight until groups of cardinality at most one, represented by their remaining bits, have been individuated.

There is accordingly a simple way to recover the original set from the tree: simply read off all branches from the root to leaves, interpreting a left going edge as a 0 and a right going edge as a 1, appending for each branch the binary string stored at the leaf. For instance, in the above tree, going left-right-right (*i.e.* reading 011), we find leaf *B* that contains the information 01, and this corresponds to the key $b = 01101$.

1.2. Operations on Tries

The recursive construction of a trie, usually represented in the form of a linked structure, closely mimics the definition given above. Once *trie* (ω) has been constructed, we can perform various operations [11, 15]:

> *query:* to determine whether u is in ω, follow a branch taking directions corresponding to the successive letters of u until a leaf is encountered, then compare with the remaining bits.
>
> *insert:* same as query; when a leaf is encountered, split it to obtain the new trie.
>
> *delete:* a dual of the insertion procedure.
>
> *union-intersection:* when sets are represented as tries, these operations can be implemented by means of the recursive definitions:

$$union\ (\psi, \vartheta) \equiv 0.\ union\ (\psi/0, \vartheta/0) \bigcup 1.\ union\ (\psi/1, \vartheta/1)$$

$$inter\ (\psi, \vartheta) \equiv 0.\ inter\ (\psi/0, \vartheta/0) \bigcup 1.\ inter\ (\psi/1, \vartheta/1),$$

with adequate initial conditions.

The costs of these operations are largely determined by the number of pointer chains followed (or bits inspected). They usually admit *inductive definitions* of a simple form over the tree structure. A prototype is the size of the trie representation measured in the number of its internal nodes, which satisfies: †

$$in \ (\omega) = \text{if } |\omega| \leq 1 \text{ then } 0 \text{ else } 1 + in \ (\omega/0) + in(\omega/1),$$

or equivalently

$$in \ (\omega) = 1 + in \ (\omega/0) + in \ (\omega/1) - \delta_{|\omega|,0} - \delta_{|\omega|,1}$$

together with the initial condition $in \ (\omega) = 0$ if $|\omega| \leq 1$.

Our purpose in what follows is to describe a method for obtaining estimates of average values of such parameters when the number of elements in the set is a fixed integer n.

1.3. Trie Indexes

In order to save storage (reduce the number of pointers used). Sussenguth [21], followed by Knuth [11, ex. 6.3.20] proposed using a sequential storage discipline whenever reaching a subfile of b or less keys. The corresponding tree which we shall call a *b-trie* thus consists of:

(i)   a skeleton tree formed with the internal nodes

(ii)   leaves containing between 0 and b records.

This idea may be used to access files on some secondary storage device. The skeleton tree then becomes the *index* or *directory* and small subfiles are stored in *pages*, with b being the page capacity determined by physical characteristics of the device (in practice b ranges

---

† *It is clear that such inductive definitions may be expressed either in terms of the left-subtree/right-subtree decomposition of trees or in terms of the equivalent decomposition of sets ω into $\omega/0$ and $\omega/1$.*

between a few tenths and a few hundreds). When used in conjunction with hashing, the resulting algorithm is exactly Larson's Dynamic Hashing method.

Finally if the index is itself too large to fit in core, it may be paged as an array that represents its embedding into a perfect tree. This algorithm constitutes the Extendible Hashing method.

Naturally, the operations described in Section 1.2 are easily adapted to such representations of files. Notice, for instance, that a query with Dynamic Hashing requires only one disk probe, Extendible Hashing which can be used even for very large files requiring only two accesses. These strategies thus guarantee an almost *direct access* to external files, whence their practical interest.

2. The Uniform Model

Our objective is to obtain estimates of expected values of a number of parameters on tries (size, path length, height, ...) as a function of the number n of elements on which the trie is built. In order to do so, we must first make precise what our probabilistic assumptions are. Following [11, 15] we retain two models.

1. *The finite key model;* Keys are to be of some fixed length s (s a non-negative integer). All sets of n elements are assumed to be equally likely. Since the number of these is:

$$b_n{}^{(s)} = \binom{2^s}{n}$$

the probability of each set is thus $(b_n{}^{(s)})^{-1}$.

2. *The infinite key model:* it is also occasionally called the Bernoulli model. Keys are assumed to be infinitely long strings of zeros and ones *i.e.* from the universe:

$$\mathbf{B}^{(\infty)} = \left\{0,1\right\}^{\infty}.$$

Alternatively, keys can be conceived for as real numbers over the real interval $[0;1]$ (the correspondence between $\mathbf{B}^{(\infty)}$ and the real interval is bijective except for a set of measure 0). The infinite key model assumes that n keys are drawn uniformly and independently over the interval $[0; 1]$.

We consider *parameters* (also called *valuations*) on sets of strings. These are here usually parameters of the trie representation of sets.

Notations: *Let $v(\omega)$ be a parameter of sets $\omega \subset \mathbf{B}^{(s)}$, $s \leq \infty$ (or of trie (ω)). We define the quantities:*

$$v_n^{(s)} = \sum_{\substack{\omega \subset \mathbf{B}^{(s)} \\ |\omega| = n}} v(\omega) \quad \text{if } s < \infty,$$

$$v_n^{(\infty)} = E[v(\omega) \mid \omega \subset \mathbf{B}^{(\infty)}, |\omega| = n]†.$$

In other words, $v_n^{(s)}$ represents the cumulated value of parameter v over all n subsets of $\mathbf{B}^{(s)}$, and $v_n^{(\infty)}$ is the expectation of random variable v on a random n-subset of $\mathbf{B}^{(\infty)}$. (Notice that one can prove that any parameter that is polynomially bounded in the size of the trie has a finite expectation under the infinite key model).

Notation: *We define the ordinary generating function of the $v_n^{(s)}$ as:*

$$v^{(s)}(x) \equiv \sum_{n=0}^{2^s} v_n^{(s)} x^n,$$

and the exponential generating function of the $v_n^{(\infty)}$ as:

$$v^{(\infty)}(x) = \sum_{n \geq 0} v_n^{(\infty)} \frac{x^n}{n!},$$

---

† *Here $E[X]$ denotes the expectation of the random variable X.*

In the sequel, we adhere to the convention of denoting parameters, corresponding cumulated values, expectations and generating functions by the same group of letters, as we have done above. We also make some use of the classical notation:

$$[x^n] f(x)$$

to represent the coefficient of x^n in the Taylor expansion of $f(x)$.

2.1. The Finite Model

In the finite model, the universe of keys is the set $\mathbf{B}^{(s)}$. The universe of sets is thus

$$\mathbf{P}^{(s)} = \left\{ \omega \subset \mathbf{B}^{(s)} \right\},$$

and the finite model consists in assuming a uniform distribution on the elements of $\mathbf{P}^{(s)}$ of cardinality n.

The definitions of additive and multiplicative valuations on tries can be translated directly into recurrence equations over generating functions as the following lemma shows:

Lemma 1 [*The additive-multiplicative translation lemma; finite model*]: *To the following schemas defining valuations on tries*

$$a(\omega) = \lambda b(\omega) \tag{i}$$

$$a(\omega) = b(\omega) + c(\omega) \tag{ii}$$

$$a(\omega) = b(\omega/0).\ c(\omega/1), \tag{iii}$$

there correspond the following relations on generating functions:

$$a^{(s)}(x) = \lambda b^{(s)}(x) \tag{i}$$

$$a^{(s)}(x) = b^{(s)}(x) + c^{(s)}(x) \tag{ii}$$

$$a^{(s)}(x) = b^{(s-1)}(x) . \; c^{(s-1)}(x) \tag{iii}$$

Proof: Relations (i) and (ii) follow from the linearity of generating functions and expectations. Relation (iii) can be established without using recurrences by writing:

$$a^{(s)}(x) = \sum_{\omega \in \mathbf{P}^{(s)}} a^{(s)}(\omega) x^{|\omega|} \tag{2.1}$$

$$= \sum_{\omega \in \mathbf{P}^{(s)}} b(\omega/0) x^{|\omega/0|} c(\omega/1) x^{|\omega/1|} \tag{2.2}$$

$$= \sum_{\omega_0 \in \mathbf{P}^{(s-1)}} b(\omega_0) x^{|\omega_0|} \sum_{\omega_1 \in \mathbf{P}^{(s-1)}} c(\omega_1) x^{|\omega_1|} \tag{2.3}$$

$$= b^{(s-1)}(x) c^{(s-1)}(x) \;\; . \tag{2.4}$$

Note that the transition from (2.2) to (2.3) results from the standard isomorphisms:

$$\mathbf{B}^{(s)} \sim (0 + 1) \, \mathbf{B}^{(s-1)} \, ,$$

$$\mathbf{P}^{(s)} \sim \mathbf{P}^{(s-1)} {}_{\mathbf{X}} \mathbf{P}^{(s-1)} \, . \quad \blacksquare$$

Lemma 2: [*The translation, lemma for initial valuations; finite model*] *: The valuations:*

$$a(\omega) = 1 \tag{i}$$

$$b(\omega) = \delta_{|\omega|,p} \tag{ii}$$

$$c(\omega) = |\omega| \tag{iii}$$

(δ *denotes the Kronecker symbol) have corresponding generating functions:*

$$a(x) = (1 + x)^{2^s} \tag{i}$$

$$b^{(s)}(x) = \binom{2^s}{p} x^p \tag{ii}$$

$$c^{(s)}(x) = 2^s x (1+x)^{2^s - 1} \tag{iii}$$

The above correspondences have a number of direct implications. For instance, if:

$$a(\omega) = b(\omega/0) \tag{2.5}$$

which we may rewrite (using 1_u to denote the valuation identically equal to 1) as

$$a(\omega) = b(\omega/0) . 1_{\omega/1} ,$$

we get:

$$a^{(s)}(x) = (1 + x)^{2^{s-1}} . b^{(s-1)}(x) \tag{2.6}$$

An important pattern in analyses is relative to parameters that are recursively defined over the tree structure. By (2.6), if a parameter v satisfies the inductive definition:

$$v(\omega) = v(\omega/0) + v(\omega/1) + w(\omega), \tag{2.7}$$

then:

$$v^{(s)}(x) = 2(1 + x)^{2^{s-1}} v^{(s-1)}(x) + w^{(s)}(x). \tag{2.8}$$

Equations of the form (2.7) are solved by iterating (or unwinding) the recurrence, and one has trivially:

Lemma 3: [*The iteration lemma for the finite model*] *The solution to the recurrence:*

$$v^{(s)}(x) = \alpha_s(x) v^{(s-1)}(x) + \beta_s(x),$$

where α, β are known and $v^{(0)}(x) = \beta_0(x)$ has the explicit solution:

$$v^{(s)}(x) = \sum_{j=0}^{s} [\beta_j(x), \prod_{k=j+1}^{s} \alpha_j(x)] \ .$$

Application of this lemma to the special case of (2.8) results in the solution:

$$v^{(s)}(x) = \sum_{j=0}^{s} 2^{s-j} w^{(j)}(x)(1+x)^{2^s-2^j} \ .$$

2.2. The Infinite Model

Our treatment of the infinite model closely follows what we have done in the preceding section. The universe of keys is now the set:

$$\mathbf{B}^{(\infty)}$$

and the universe of sets is:

$$\mathbf{P}^{(\infty)} = \left\{ \omega \subset \mathbf{B}^{(\infty)} \mid \omega \ \text{finite} \right\}.$$

The basic property here is that for a random set ω of n elements, the probability that the size of $\omega/0$ be equal to k and the size of $\omega/1$ be equal to $n-k$ is simply the Bernoulli probability:

$$\beta_{n.k} = \frac{1}{2^n} \binom{n}{k} \ .$$

We have:

Lemma 4: [*The additive-multiplicative translation lemma; infinite model*]: *If valuations on tries satisfy the relations:*

$$a(\omega) = \lambda b(\omega) \tag{i}$$

$$a(\omega) = b(\omega) + c(\omega) \tag{ii}$$

$$a(\omega) = b(\omega/0).c(\omega/1), \tag{iii}$$

then the corresponding generating functions are related by:

$$a^{(\infty)}(x) = \lambda b^{(\infty)}(x) \tag{i}$$

$$a^{(\infty)}(x) = b^{(\infty)}(x) + c^{(\infty)}(x) \tag{ii}$$

$$a^{(\infty)}(x) = b^{(\infty)}(\frac{x}{2}).c^{(\infty)}(\frac{x}{2}). \tag{iii}$$

Proof: Again (i) and (ii) are trivial. Relation (iii) is proven by:

$$a^{(\infty)}(x) = \sum_{n \geq 0} a_n^{(\infty)} \frac{x^n}{n!} = \sum_{n, k \geq 0} \beta_{n,k} b^k c_{n-k} \frac{x^n}{n!}$$

$$= \sum_{n_1 \geq 0} b_{n_1} \frac{x^{n_1}}{n_1!} \sum_{n_2 \geq 0} c_{n_2} \frac{x^{n_2}}{n_2!}$$

$$= b^{(\infty)}(\frac{x}{2}).c^{(\infty)}(\frac{x}{2}).$$

The product form comes from the fact that when ω is a random n-subset of $\mathbf{B}^{(\infty)}$, then if $\omega/0$ is conditioned to be of cardinality k, it is a random k-subset of $\mathbf{B}^{(\infty)}$. ∎

Lemma 5 [*The translation lemma for initial valuations; infinite model*] *: The valuations:*

$$a(\omega) = 1 \tag{i}$$

$$b(\omega) = \delta_{|\omega|,p} \tag{ii}$$

$$c(\omega) = |\omega| \tag{iii}$$

have corresponding generating functions:

$$a^{(\infty)}(x) = e^x \qquad\qquad\qquad (i)$$

$$b^{(\infty)}(x) = \frac{x^p}{p!} \qquad\qquad\qquad (ii)$$

$$c^{(\infty)}(x) = x e^x. \qquad\qquad\qquad (iii)$$

We again have the important special cases corresponding to (2.5), (2.7). If

$$a(\omega) = b(\omega/0)$$

then

$$a^{(\infty)}(x) = e^{x/2} b^{(\infty)}\left(\frac{x}{2}\right); \qquad\qquad (2.9)$$

similarly, if $v(\omega)$ is a recursively defined parameter:

$$v(\omega) = v(\omega/0) + v(\omega/1) + w(\omega)$$

then

$$v^{(\infty)}(x) = 2e^{x/2} v^{(\infty)}\left(\frac{x}{2}\right) + w^{(\infty)}(x). \qquad\qquad (2.10)$$

Equations of the form (2.10) may be solved by iteration, and we have in analogy to Lemma 3:

Lemma 6 [*The iteration lemma; infinite case*] : *Let $\alpha(x)$ and $\beta(x)$ be two entire functions such that $\alpha(0) = c$ and $\beta(x) = 0(x^d)$ as $x \to 0$. The difference equation:*

$$f(x) = \alpha(x) f\left(\frac{x}{2}\right) + \beta(x)$$

where f is the unknown function and α, β satisfy the "contraction condition":

$$c 2^{-d} < 1,$$

with the initial conditions on $f(x)$:

$$f(0) = f'(0) = f''(0) = \cdots = f^{(d-1)}(0) = 0$$

has a unique entire solution given by:

$$f(x) = \sum_{j \geq 0} [\beta(\frac{x}{2^j}) \prod_{k=0}^{j-1} \alpha(\frac{x}{2^k})].$$

Proof: Iterating the basic equation, one gets:

$$f(x) = \beta(x) + \alpha(x) f(\frac{x}{2})$$

$$= \beta(x) + \alpha(x)\beta(\frac{x}{2}) + \alpha(x)\alpha(\frac{x}{2}) f(\frac{x}{4})$$

$$= \cdots$$

The initial conditions together with the contraction condition ensure the convergence of the infinite sum that one obtains in the limit. ∎

We have again an importan special case corresponding to (2.10) where two equivalent expressions can be derived.

Lemma 7 [*Iteration Lemma for the infinite model; special case*] *The difference equation:*

$$f(x) = ce^{x/2} f(\frac{x}{2}) + \beta(x)$$

admits provided the initial condition and the contraction conditions of Lemma 6 are satisfied the solution:

$$f(x) = \sum_{j \geq 0} c^j \beta(\frac{x}{2^j}) e^{x(1-\frac{1}{2^j})}. \tag{i}$$

Alternatively, the Taylor expansion of f(x):

$$f(x) = \sum_{n \geq 0} f_n \frac{x^n}{n!}$$

can be obtained as:

$$f_n = \sum_k \binom{n}{k} \frac{\beta_k^*}{1-c \, 2^{-k}}, \text{ where } \beta_k^* = k! \, [x^k] e^{-x} \beta(x).$$ (ii)

Proof: Part (i) is a direct application of Lemma 6. For part (ii), we set $f^*(x) = e^{-x} f(x)$; $f^*(x)$ satisfies:

$$f^*(x) = c \, f^* \left(\frac{x}{2} \right) + \beta^*(x).$$ (2.11)

Identifying coefficients of x^n in (2.11) gives the relation $f_n^* = c \, 2^{-n} f_n^* + \beta_n^*$. Relation (ii) then follows since the coefficients of $f(x)$ are convolutions of those of $f^*(x)$ by e^x. ∎

Notice that in applications, the initial condition on $f(x)$ can be bypassed by subtracting from f adequate combinations of functions of the form: $x^m e^{px}$.

The reader may compare this approach to the treatment of recurrences occurring in the analysis of trie parameters via the use of *binomial transforms* in [11, ex. 5.2.2.36-38].

Notice, as a final remark in this section, the following relation between the finite and the infinite models [7]:

$$v^{(\infty)}(x) = \lim_{s \to \infty} v^{(s)} \left(\frac{x}{2^s} \right),$$

which is clear on the coefficients and explains many of the analogies between the two models.

3. Alternative Models

The way taken in Section 2 which allows for a systematic *translation mechanism* from parameter definitions to generating functions may be extended to a diversity of models. Since the proof techniques in each case differ only little from what we have encountered, we only briefly sketch here the kind of results that can be attained.

3.1. Multiway Tries

In many applications, one may wish to take advantage of the decomposition of records into characters or bytes instead of bits. The resulting trees have then a branching degree corresponding to the cardinality of the alphabet which is an integer m, $m \geq 2$.

The definition of multiway tries mimics closely that of binary tries; if the alphabet is assimilated to the integer interval $[1.. m]$, we consider the sets:

$$\mathbf{M}^{(s)} = [1.. m]^s; \quad \mathbf{M}^{(\infty)} = [1.. m]^\infty,$$

and tries are now defined recursively via the decomposition:

$$\omega \equiv 1.(\omega/1) \bigcup 2.(\omega/2) \bigcup \ldots \bigcup m.(\omega/m),$$

for any $\omega \subset \mathbf{M}^{(s)}$, with $\omega/j \subset \mathbf{M}^{(s-1)}$. Sum and product rules remain valid as before (with m-ary products if a valuation is a product of m valuations on subtrees).

In the finite case, the generating function describing the universe of all subsets of $\mathbf{M}^{(s)}$ becomes:

$$\sum_{\omega \subset \mathbf{M}^{(s)}} x^{|\omega|} = (1+x)^{m^s}. \tag{3.1}$$

In particular, if

$$v(\omega) = w(\omega/1), \tag{3.2}$$

we have:

$$v^{(s)}(x) = w^{(s-1)}(x)((1+x)^{m^{s-1}})^{m-1}$$

$$\equiv w^{(s-1)}(x)(1+x)^{m^s - m^{s-1}}. \tag{3.3}$$

In the infinite case, for instance, (3.2) leads to

$$v^{(\infty)}(x) = w^{(\infty)} \left(\frac{x}{m} \right) (e^{x/m})^{m-1}$$

$$= w^{(\infty)} \left(\frac{x}{m} \right) e^{x\left(1 - \frac{1}{m}\right)} . \tag{3.4}$$

3.2. Biased Bits

To model more closely some applications, as for instance when tries are built out of textual data, one also consider *non uniform probability distributions* on bits or characters of strings. We shall study here the infinite model only. Starting with the binary case, the model assumes that bits of keys are taken independently from some discrete distribution:

$$Pr(0\text{--}bit) = p; \quad Pr(1\text{--}bit) = q \equiv 1-p. \tag{3.5}$$

In other terms, for $u \in \mathbf{B}^{(\infty)}$, we have:

$$Pr\left(|u_1 u_2 \ldots u_l|_0\right) = \binom{l}{k} p^k q^{n-k}.$$

($|u|_0$ denotes the number of zeros in u).

Additive properties of generating functions still hold. The main difference lies in multiplicative valuations, for which:

$$v(\omega) = w(\omega/0)\, t\,(\omega/1) \tag{3.6}$$

translates into:

$$v^{(\infty)}(x) = w^{(\infty)}(px).t^{(\infty)}(qx). \tag{3.7}$$

This biased model also extends easily to m-ary tries, as we have been considering in Section 3.1. If the probability distribution on an m-ary alphabet is $(p_1, p_2, \cdots, p_m)$ with $\sum p_i = 1$, then

$$v(\omega) = w_1(\omega/1). \; w_2(\omega/2) \cdots w_m(\omega/m) \qquad (3.8)$$

translates into:

$$v^{(\infty)}(x) = w_1{}^{(\infty)}(p_1 x). \; w_2{}^{(\infty)}(p_2 x) \cdots, \; w_m{}^{(\infty)}(p_m x). \qquad (3.9)$$

3.3. Allowing Repetitions

The definition of tries associated to *sets* of binary_or other_sequences can also be extended to *multisets* where elements may appear repeated several times. In order to do so, we only need to allow leaves to contain several elements that are identical. In practice the situation occurs for instance when constructing tries on a single field of composite records. Although records are usually all distinct, some values of a specified field are likely to occur several times (many people live in New-York City!).

Our universe of "files" has now become in the binary case the family $\mathbf{Q}^{(s)}$ of all multisets over $\mathbf{B}^{(s)}$ which, using notations from formal language theory may be rewritten as:

$$\mathbf{Q}^{(s)} = \prod_{\alpha \in \mathbf{M}^{(s)}} \alpha^* \qquad (3.10)$$

with:

$$\alpha^* = \phi + \alpha + \alpha^2 + \alpha^3 + \cdots. \qquad (3.11)$$

Taking as a measure of the size $|\,.\,|$ of a multiset the number of its elements counted with their multiplicities, the generating function that describes the universe of multisets $\mathbf{Q}^{(s)}$ is found to be:

$$\sum_{\omega \in \mathbf{Q}^{(s)}} x^{|\omega|} = \left(\frac{1}{1-x}\right)^{2^s}. \qquad (3.12)$$

Notice in passing the formal analogy between definitions (3.10), (3.11) and equation (3.12).

Equation (3.12) is also consistent with the obvious counting result:

$$\mathrm{card}\left\{\omega \in \mathbf{Q}^{(s)} \mid |\omega| = n\right\} = \binom{2^s + n - 1}{n}. \tag{3.13}$$

Sum and product rules again apply and it is only in subtree valuations that the form (3.12) of the "universal" polynomial has to be taken into account.

For instance, if

$$v(\omega) = w(\omega/0),$$

then under this model:

$$v^{(s)}(x) = w^{(s-1)}(x)(1-x)^{-(2^{s-1})}.$$

3.4. Several Sets

Set-theoretic operations like union, intersection, ... take as arguments several sets. In order to analyse them in the average case, we should therefore enter the sizes of the arguments as parameters. Restricting here the discussion to the case of *two* sets, we consider valuations of the form:

$$v : \mathbf{P}^{(s)} \times \mathbf{P}^{(s)} \to \mathbf{R}.$$

The cumulated values of v:

$$v_{m,n}^{(s)} = \sum_{\substack{\xi,\eta \in \mathbf{P}^{(s)} \\ |\xi|=m, |\eta|=n}} v(\xi,\eta)$$

can be attained through the *bivariate* generating functions:

$$v^{(s)}(x,y) \equiv \sum_{m,n} v_{m,n}^{(s)} x^m y^n = \sum_{\xi,\eta \in \mathbf{P}^{(s)}} v(\xi,\eta) x^{|\xi|} y^{|\eta|}.$$

As we shall see in Section 4.2, a relation like:

$$v(\xi, \eta) = w(\xi/0, \eta/0)$$

translates into:

$$v^{(s)}(x, y) = (1+x)^{2^{s-1}}(1+y)^{2^{s-1}} w^{(s-1)}(x, y).$$

Proof techniques are highly simplified if one uses the way taken in Section 2.1. This permits us in particular to give a detailed analysis of trie union and trie intersection.

3.5. The Poisson Model

The Poisson model has been used to obtain expressions that are sometimes easier to handle than corresponding expressions under the Bernoulli (infinite key) model. A typical example is the treatment of directory size in Extendible Hashing [2] that will be discussed in Section 4.3.

Under this model, the number N of elements on which a trie is constructed is a Poisson distributed random variable with average n (n being a parameter). The keys themselves are uniformly distributed over the real interval $[0; 1]$. We have:

Lemma 8: *If a parameter v has under the Bernoulli model for n keys an expected value $v_n^{(\infty)}$, then under a Poisson model of parameter v, its expectation satisfies:*

$$v_v^p = \sum_{k=0}^{\infty} e^{-v} \frac{v^k}{k!} v_k^{(\infty)} = e^{-v} v^{(\infty)}(v).$$

Lemma 8 is a trivial consequence of the form of the Poisson probability distribution. It shows that the *values* at positive real points of *generating functions* under the Bernoulli model are directly related to the *Poisson expectations*. Thus Lemmas 4,5 translate *verbatim* into schemes that permit to determine from the shape of valuations the expected values of trie parameters under the Poisson model.

4. Applications to Digital Search

4.1. Simple Operations on Tries

We analyse here the storage efficiency of tries (and of some of their variants), as well as the time cost of a basic search. Our aim in this section is to provide a uniform framework for a number of results that are to be found in [11]. (See in particular Sect. 5.2.2 and ex. 5.2.2.36-38; Sect. 6.3 and ex. 6.3.20.31-34)

Multiway Tries

Our first analysis is relative to the storage occupation of multiway tries. In the case of an alphabet of cardinality m, assimilated to the integer interval $[1..m]$, the determinant parameter is the number of internal nodes of the trie. This parameter admits, as we have seen, the inductive definition:

$$in(\omega) = in(\omega/1) + in(\omega/2) + \cdots + in(\omega/m) + 1 - \delta_{|\omega|.0} - \delta_{|\omega|.1} \qquad (4.1)$$

for $\omega \subset \mathbf{B}^{(s)}$, with $s \geq 1$ or $s = \infty$. For $\omega \subset \mathbf{B}^{(0)}$, we have $in(\omega) = 0$.

In the finite case, we find from Section 3.1 the recurrence relation:

$$in^{(s)}(x) = (1+x)^{m^s} - 1 - m^s x + m(1+x)^{m^s - m^{s-1}} in^{(s-1)}(x) \qquad (4.2)$$

for $s \geq 1$, with $in^{(0)}(x) = 0$. In the infinite case, we get a difference equation:

$$in^{(\infty)}(x) = m\, e^{x(1 - \frac{1}{m})} in^{(\infty)}(\frac{x}{m}) + e^x - 1 - x. \qquad (4.3)$$

Equation (4.2) is readily solved by unwinding the recurrence, and we obtain:

$$in^{(s)}(x) = \sum_{j=1}^{s} m^{s-j}(1+x)^{m^s - m^j}[(1+x)^{m^j} - 1 - m^j x]. \qquad (4.4)$$

Taking Taylor coefficients of formula (4.4), we obtain the explicit form of the total number of nodes in all tries formed with n distinct keys of length s over an m-ary alphabet:

$$in_n^{(s)} = \sum_{j=1}^{s} m^{s-j} \binom{m^s}{n} - m^{s-j} \binom{m^s - m^j}{n} - m^s \binom{m^s - m^j}{n-1}$$

Solving under the infinite model is even simpler. By the methods of Section 2, we find for the exponential generating function of expected values of the number of internal nodes of tries the relation:

$$in^{(\infty)}(x) = \sum_{k \geq 0} m^k [e^x - e^{x(1-\frac{1}{m^k})} - \frac{x}{m^k} e^{x(1-\frac{1}{m^k})}]. \tag{4.5}$$

Taking again Taylor coefficients in (4.5), we find for the expectation of *in* under the infinite key model the form:

$$in_n^{(\infty)} = \sum_{k \geq 0} m^k [1 - (1 - \frac{1}{m^k})^n - \frac{n}{m^k}(1 - \frac{1}{m^k})^{n-1}]. \tag{4.6}$$

Path length in multiway tries can be analysed in very similar terms. From the definition:

$$pl(\omega) = pl(\omega/1) + pl(\omega/2) + \cdots + pl(\omega/m) + |\omega| - \delta_{|\omega|,1}, \tag{4.7}$$

we find the equations corresponding to the finite and infinite models:

$$pl^{(s)}(x) = m.pl^{(s-1)}(x)(1+x)^{m^s - m^{s-1}} + m^s x[(1+x)^{m^s-1} - 1] \tag{4.8}$$

$$pl^{(\infty)}(x) = m.e^{x(1-\frac{1}{m})} pl^{(\infty)}(\frac{x}{m}) + x(e^x - 1). \tag{4.9}$$

Solutions may be obtained as before, and summarising these analyses, we find [11]:

Theorem A: *The expectation of the number of nodes in an m-ary trie constructed with n keys is:*

$$in_n^{(s)} = \sum_{j=1}^{s} m^{s\cdot j} \binom{m^s}{n} - m^{s\cdot j} \binom{m^s - m^j}{n} - m^s \binom{m^s - m^j}{n-1},$$

$$in_n^{(\infty)} = \sum_{k \geq 0} m^k [1 - (1 - \frac{1}{m^k})^n - \frac{n}{m^k}(1 - \frac{1}{m^k})^{n-1}].$$

The expected value of path length is:

$$pl_n^{(s)} = m^s \sum_{j=1}^{s} [\binom{m^s - 1}{n--1} - \binom{m^s - m^j}{n-1}]$$

$$pl_n^{(\infty)} = n \sum_{j \geq 0} [1 - (1 - \frac{1}{m^j})^{n-1}].$$

In particular, the expected cost, measured in the number of bit inspections, of a positive search is

$$\frac{1}{n\binom{m^s}{n}} pl_n^{(s)}, \quad \frac{1}{n} pl_n^{(\infty)}.$$

Binary Representations of Multiway Tries

In the case of multiway tries, the asymptotic analysis of the number of nodes reveals that storing a file of n elements necessitates about $\frac{m}{\log m} n$ pointers, which may be quite expensive when m is large (many such pointers toward the low levels in the tree are likely to be null). For that reason, it may prove necessary to use a binary representation of tries, where each node is linked to its leftmost son and its immediate right brother. The price to be paid is an increased time cost, since access to subtrees is now done in a sequential way. Such a representation is displayed in Figure 2.

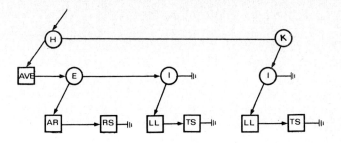

Figure 2: The binary tree encoding of a trie associated to the file { HAVE, HEAR, HERS, HILL, HITS, KILL, KITS }.

There are several conceivable implementations of this structure. In the one shown in Figure 2, an internal node of the original trie necessitates two pointers while external nodes only use one pointer. The problem of the storage occupation of this structure is thus solved by our previous analyses. We propose here to analyse the cost of a positive search under the infinite model.

Let $bp(\omega)$ be the total number of pointers traversed when searching all keys in the binary tree representation of *trie* (ω). An inductive definition of this quantity is obtained by observing that the cost of accessing the subtree corresponding to ω/k is equal to 1 plus the number of non-empty subsets ω/j, for $0 \leq j < k$. Hence: †

$$bp(\omega) = |\omega| + \sum_{k=1}^{m} [bp(\omega/k) + |\omega/k| \sum_{j=0}^{k-1} \chi(|\omega/j| \geq 1)] \quad (4.10)$$

From there, the translation to generating functions is immediate, and, for the infinite model, we have:

† *We let $\chi(P)$ denote the function whose value is 0 if P is false and 1 if P is true.*

$$bp^{(\infty)}(x) = x(e^x-1)+me^{x(1-\frac{1}{m})} bp^{(\infty)}(\frac{x}{m})+\frac{m(m-1)}{2}(e^{\frac{x}{m}}-1) \cdot$$

$$\cdot \; \frac{x}{m}e^{\frac{x}{m}}(e^{\frac{x}{m}})^{m-2} =$$

$$x[\frac{m+1}{2}e^x - \frac{m-1}{2}e^{x(1-\frac{1}{m})}-1] + me^{x(1-\frac{1}{m})} bp^{(\infty)}(\frac{x}{m}). \quad (4.11)$$

Solving, we obtain:

Theorem B: *The average number of pointers of a binary tree representation of a multiway trie of n elements is:*

$$2in_n^{(\infty)} +n.$$

The average number of pointers followed in a positive search is

$\frac{1}{n} bp_n^{(\infty)}$, *where:*

$$bp_n^{(\infty)} = n \sum_{k\geq 0} [\frac{m+1}{2} - \frac{m-1}{2}(1-\frac{1}{m^{k+1}})^{n-1} -(1-\frac{1}{m^k})^{n-1}].$$

Patricia Trees

Patricia Trees are a compact representation of tries due to R. Morrison in which one-way branching is avoided by means of skip fields. Our description follows [11, pp. 497-498]. We propose here to analyse under the infinite model the cost of both positive and negative queries.

The access cost of a leaf in a Patricia tree is therefore exactly the number of binary nodes traversed in the corresponding trie. Thus the cost of a positive query in a tree of n elements is $\frac{1}{n}$ times the "path length through binary nodes" in the underlying trie. This modified path length is defined by the recursion:

$$ppl(\omega)=\begin{cases} ppl(\omega/0)+ppl(\omega/1)+|\omega/0|+|\omega/1| & \text{if } |\omega/0|.|\omega/1|\neq0 \\ ppl(\omega/0) & \text{if } |\omega/1| = 0 \\ ppl(\omega/1) & \text{if } |\omega/0| = 0 \end{cases}$$

This definition can be trivially rephrased as an additive-multiplicative combination of standard valuations. From there, we obtain the equation (terms correspond to those of the above definition):

$$ppl^{(\infty)}(x) = 2(e^{x/2}-1)ppl^{(\infty)}(\frac{x}{2})+2\,e^{x/2}\,\frac{x}{2}\,(e^{x/2}-1) +$$

$$+2ppl^{(\infty)}(\frac{x}{2}), \tag{4.12}$$

which, after simplification gives:

$$ppl^{(\infty)}(x) = 2e^{x/2}ppl^{(\infty)}(\frac{x}{2})+x(e^x - e^{x/2}). \tag{4.13}$$

The case of a negative search leads to a difference equation of a new type. Let $pns(\omega)$ denote the expected cost of a random (*i.e.* negative with probability 1) search in the Patricia tree built on ω. This cost, again measured in the number of pointers followed, has the definition:

$$pns(\omega) = \begin{cases} 1/2\,pns\,(\omega/0)+1/2\,pns(\omega/1)+1 & \text{if } |\omega/0|.|\omega/1|\neq0 \\ pns\,(\omega/0) & \text{if } |\omega/1| = 0 \\ pns\,(\omega/1) & \text{if } |\omega/0| = 0 \end{cases}$$

This definition can be translated as before into:

$$pns^{(\infty)}(x) = (e^{x/2}-1)pns^{(\infty)}(\frac{x}{2})+(e^{x/2}-1)^2 +2pns^{(\infty)}(\frac{x}{2}).$$

$$= (1+e^{x/2})pns^{(\infty)}(\frac{x}{2})+(e^{x/2}-1)^2. \tag{4.14}$$

This equation can be solved by the iteration method described in Section 2:

$$pns^{(\infty)}(x) = \sum_{k \geq 0} (e^{2^{\frac{x}{k+1}}} - 1)^2 \prod_{j=0}^{k-1} (1+e^{2^j}),$$

which using the identity:

$$(1+q)(1+q^2)(1+q^4) \cdots (1+q^{2^{j-1}}) = \frac{(1-q^{2^j})}{(1-q)}$$

yields the explicit form:

$$pns^{(\infty)}(x) = \sum_{k \geq 1} \frac{1-e^{2^{\frac{x}{k}}}}{1+e^{2^{\frac{x}{k}}}} (1-e^x). \tag{4.15}$$

To extract the Taylor coefficients of $pns^{(\infty)}(x)$ from (4.15), our route now follows that of Knuth [11, ex. 6.3.34.a]. We notice that for a Taylor series:

$$f(x) = \sum_{n \geq 1} f_n x^n,$$

one obtains by expanding then aggregating the coefficients of x^n:

$$\sum_{k \geq 1} f(\frac{x}{2^k}) = \sum_{n \geq 1} f_n \frac{x^n}{2^n-1}.$$

Thus, from the standard definition of the Bernoulli numbers:

$$\sum_{n \geq 1} B_{n+1} \frac{x^n}{(n+1)!} (2^{n+1}-1) = \tanh(\frac{x}{2}) = \frac{e^x-1}{e^x+1},$$

we can expand the factor of $(1-e^x)$ in (4.15). The coefficients of the function *pns* are finally obtained by multiplying the expansion so obtained by the expansion of $(1-e^x)$. We have thus proved [11]:

Theorem C: *The expected costs of a positive and a negative search in a Patricia tree with n keys are:*

$$\frac{1}{n} pl_n^{(\infty)} - 1,$$

$$pns_n^{(\infty)} = \frac{2}{n+1} - 3 + 2 \sum_{k \geq 1} \frac{1}{k+1} \binom{n}{k} \frac{B_{k+1}}{2^k - 1}.$$

4.2. Set Theoretic Operations on Tries

Union and intersection can be efficiently performed on trie representations of sets. As we saw in section 1.2, the algorithms are based on a simultaneous traversal of tries. We propose here to give a precise analysis of *trie intersection*. Our results improve on Trabb Pardo's [15] who only had an approximate analysis.

The parameters of the analysis are the cardinalities m, n of both sets whose intersection is to be computed *and* the cardinality of their intersection k. This way of proceeding follows Trabb Pardo's approach and is motivated by the fact that random sets tend to have very small intersections. Thus taking also the size of the intersection into account yields more informative results.

Our statistical model thus assumes all pairs of sets:

$$I_{m,n,k}^{(s)} = \left\{ (\xi, \eta) \mid \xi, \eta \subset B^{(s)}, \; |\xi| = m|, \; |\eta| = n, \; |\xi \cap \eta| = k \right\}$$

to be equally likely. By considering $I_{m,n}^{(s)} = \bigcup_k I_{m,n,k}^{(s)}$ *i.e.* summing our expressions over k, our results give the analysis of trie intersection performed on random sets of cardinalities m,n.

The intersection algorithm is obtained from the recursive definition of intersection:

$$inter(\xi, \eta) = 0.inter(\xi/0, \eta/0) \bigcup 1.inter(\xi/1, \eta/1).$$

with the initial conditions:

if $|\xi| = 0$ *or* $|\eta| = 0$ then report the empty set as intersection;

if $|\xi| = 1$ then search for ξ in η;

if $|\eta| = 1$ then search for η in ξ.

The *cost* of the intersection will be taken here to be the number of nodes traversed simultaneously in both tries. Extension to more complex cost measures is also possible by our methods. This cost admits an inductive definition that closely reflects the structure of the procedure:

$$ti(\xi, \eta) = 1 + ti(\xi/0, \eta/0) + ti(\xi/1, \eta/1) - \chi(|\xi| \leq 1 \text{ or } |\eta| \leq 1).$$

The first problem we encounter is to determine the number of input configurations, *i.e.* the quantity

$$I^{(s)}_{m,n,k} = card(\mathbf{I}^{(s)}_{m,n,k}).$$

Let us define the function:

$$I^{(s)}(x, y, t) = \sum_{\xi, \eta \subset \mathbf{B}^{(s)}} x^{|\xi|} y^{|\eta|} t^{|\xi \cap \eta|};$$

we have:

$$I^{(s)}_{m,n,k} = [x^m y^n t^k] I^{(s)}(x, y, t).$$

To determine the quantity $I^{(s)}(x, y, t)$, we apply the techniques of Sections 2,3.

We write:

$$I^{(s)}(x,y,t) = \sum_{\xi, \eta \subset \mathbf{B}^{(s)}} x^{|\xi/0| + |\xi/1|} y^{|\eta/0| + |\eta/1|} t^{|\xi/0 \cap \eta/0| + |\xi/1 \cap \eta/1|}$$

$$= \sum_{\xi_0, \eta_0, \xi_1, \eta_1 \subset \mathbf{B}^{(s-1)}} x^{|\xi_0| + |\xi_1|} y^{|\eta_0| + |\eta_1|} t^{|\xi_0 \cap \eta_0| + |\xi_1 \cap \eta_1|}$$

$$= [I^{(s-1)}(x, y, t)]^2.$$

Since we have the initial value:

$$I^{(0)}(x,\ y,t) = (1+x+y+txy)\ ,$$

we get:

$$I^{(s)}(x,\ y,\ t) = (1+x+y+txy)^{2^s}. \tag{4.16}$$

The same process can be applied to multiplicative valuations over sub-trees. If

$$v(\xi,\ \eta) = w(\xi/0,\ \eta/0) \tag{4.17}$$

we find

$$v^{(s)}(x,\ y,\ t) = (1+x+y+txy)^{2^{s-1}}\ w^{(s-1)}(x,\ y,\ t). \tag{4.18}$$

Applying this paradigm (4.17)-(4.18) to the equation defining ti, we have:

$$ti^{(s)}(x,\ y,\ t) = (1+x+y+txy)^{2^{s-1}}\ ti^{(s-1)}(x,\ y,\ t)$$

$$+(1+x+y+txy)^{2^s} - X^{(s)}(x,\ y,\ t), \tag{4.19}$$

where
$$X^{(s)}(x,\ y,\ t) = \sum_{\substack{\xi,\eta \subset \mathbf{B}^{(s)} \\ |\xi|\leq 1, |\eta|\leq 1}} x^{|\xi|} y^{|\eta|} t^{|\xi \cap \eta|}. \tag{4.20}$$

Determining $X^{(s)}(x,\ y,\ t)$ is routinely obtained by considering all possible cases, and we find:

$$X^{(s)}(x,\ y,\ t) = 2^s(1+x+y+txy)\ [(1+x)^{2^s-1}+(1+y)^{2^s-1}-1]$$

$$-(2^s-1)\ [(1+x)^{2^s}+(1+y)^{2^s}-1-2^s\ xy]. \tag{4.21}$$

Whence solving by means of Lemma 3 the explicit form:

$$ti^{(s)}(x, y, t) = (2^s-1)(1+x+y+txy)^{2^s} -$$

$$- \sum_{j=1}^{s} 2^{s-j}(1+x+y+txy)^{2^s-2^j} X^{(j)}(x, y, t). \tag{4.22}$$

There now remains the task of extracting coefficients of the polynomials that appear in equation (4.22). To that purpose, we define:

$$I_{m,n,k}[\alpha, \gamma] = [x^m y^n t^k](1+x)^\alpha(1+x+y+txy)^\gamma \tag{4.23}$$

$$J_{m,n,k}[\beta, \gamma] = [x^m y^n t^k](1+y)^\beta(1+x+y+txy)^\gamma.$$

Expanding first in t then in x and y the polynomials of (4.23), we immediately find:

$$I_{m,n,k}[\alpha, \gamma] = \binom{\gamma}{k}\binom{\gamma-k}{n-k}\binom{\gamma+\alpha-n}{m-k}$$

$$J_{m,n,k}[\beta, \gamma] = \binom{\gamma}{k}\binom{\gamma-k}{m-k}\binom{\gamma+\beta-m}{n-k} \tag{4.24}$$

The quantities $I_{m,n,k}[\alpha, \gamma]$ generalise the $I^{(s)}_{m,n,k}$. In particular, from (4.24), we have:

$$I^{(s)}_{m,n,k} = I_{m,n,k}[0,2^s] = J_{m,n,k}[0,2^s] = \binom{2^s}{k}\binom{2^s-k}{m-k}\binom{2^s-m}{n-k}, \tag{4.25}$$

a fact which could have been deduced by direct reasoning.

Using (4.23), (4.24) in (4.22), we finally obtain:

Theorem D: *The cumulated cost of the trie intersection algorithm applied to two sets of cardinalities m and n whose intersection has cardinality k is:*

$$ti^{(s)}_{m,n,k} = (2^s-1) I_{m,n,k}[0,2^s]$$

$$-2^s \sum_{j=1}^{s} \left\{ I_{m,n,k}[2^j-1,2^s-2^j+1]+J_{m,n,k}[2^j-1, 2^s-2^j+1] \right.$$

$$-I_{m,n,k}[0,2^s-2^j+1]+(2^j-1)\,I_{m-1,n-1,k}[0,2^s-2^j]\Big|$$

$$+\sum_{j=1}^{s}\,(2^s-2^j)\Big\{I_{m,n,k}[2^s,\,2^s-2^j]+J_{m,n,k}[2^s,\,2^s-2^j]-$$

$$-\,I_{m,n,k}[0,2^s-2^j]\Big\}\,.$$

The expected cost, assuming a uniform distribution over $\mathbf{I}^{(s)}_{m,n,k}$ *is:*

$$\frac{1}{I_{m,n,k}[0,2^s]}\,ti^{(s)}_{m,n,k}\,.$$

One could analyse in a similar way the cost of trie union as well as take into account the cost of other operations (pointer traversal, bit inspections, tests...).

4.3. *Trie Indexes*

We propose to analyse here the main parameters of trie indexes, when the keys are infinite.

We shall first evaluate the number of pages necessary to store records of the file; this quantity satisfies the recurrence:

$$page(\omega) = page\,(\omega/0)+page\,(\omega/1)-\chi(|\omega|\le b).$$

with the initial conditions: *page* $(\omega) = 1$ if $|\omega|\le b$. This is an additive valuation on tries. It is amenable to the techniques previously described, and we find for the corresponding generating function the difference equation:

$$page^{(\infty)}(x) = 2e^{x/2}\,page^{(\infty)}(\frac{x}{2}) - e_b(x), \tag{4.26}$$

where

$$e_b(x) = \sum_{j=0}^{b} \frac{x^j}{j!}.$$

Equation (4.26) does not satisfy the contraction condition of Lemma 6. However, the auxiliary function $\psi(x) = page^{(\infty)}(x) - e^x$ does. Function ψ is defined by a variant of equation (4.26):

$$\psi(x) = 2e^{x/2} \psi(\frac{x}{2}) + e^x - e_b(x),$$

which can be solved by iteration leading to an explicit form of $page^{(\infty)}$:

$$page^{(\infty)}(x) = e^x + \sum_{k \geq 0} 2^k e^{x(1-2^{-k})} (e^{x 2^{-k}} - e_b(x 2^{-k})) \qquad (4.27)$$

from which Taylor coefficients can be extracted.

The distribution of page occupation may be analysed in a similar manner. Let $page_p(\omega)$ denote the number of pages containing p elements. This quantity can be defined by:

$$page_p(\omega) = [page_p(\omega/0) + page_p(\omega/1)]\chi(|\omega| > b) + \delta_{|\omega|, p}. \qquad (4.28)$$

Under this form, equation (4.28) does not fit directly into the schemes we have previously introduced. It can however be brought under the standard form of an additive-multiplicative valuation if we rewrite it as:

$$page_p(\omega) = page_p(\omega/0) + page_p(\omega/1) \qquad (4.29)$$

$$+\chi(|\omega| = p) - \chi(|\omega/0| = p).\chi(|\omega/1| \leq b - p) -$$

$$-\chi(|\omega/1| = p).\chi(|\omega/0| \leq b - p).$$

From (4.29) follows the equation:

$$page_p^{(\infty)}(x) = 2e^{x/2} page_p^{(\infty)}(\frac{x}{2}) + \frac{x^p}{p!} - 2^{1-p}\frac{x^p}{p!}e_{b-p}(\frac{x}{2}). \quad (4.30)$$

This equation can be solved as before and one finds:

Theorem E: *The expected number of pages in a binary trie index, when the page capacity is b is given by:*

$$page_n^{(\infty)} = 1 + \sum_{k=0}^{\infty} \quad (4.31)$$

$$2^k[1-(1-\frac{1}{2^k})^n - \binom{n}{1}(\frac{1}{2^k})(1-\frac{1}{2^k})^{n-1} \cdots -$$

$$-\binom{n}{b}(\frac{1}{2^k})^b(1-\frac{1}{2^k})^{n-b}]$$

The expected number of pages containing p elements is:

$$page_{p,n}^{(\infty)} = \delta_{n,p} + \binom{n}{p}\sum_{k\geq 1} 2^{(1-p)k}[(1-\frac{1}{2^k})^{n-p}- \quad (4.32)$$

$$(1-\frac{1}{2^{k+1}})^{n-p} - \binom{n-p}{1}(\frac{1}{2^k})(1-\frac{1}{2^{k+1}})^{n-p-1} - \cdots -$$

$$-\binom{n-p}{b-p}(\frac{1}{2^k})^{b-p}(1-\frac{1}{2^{k+1}})^{n-b}].$$

We now proceed with the evaluation of the depth of b-tries wich is related to the size of the implementation of the index in extendible hashing. To that purpose, we introduce the *characteristic variables* $p_k^{(\infty)}(\omega)$, defined by:

$$p_k^{(\infty)}(\omega) = \begin{cases} 1 \text{ if the height of } \omega \text{ is at most } k \\ \\ 0 \text{ otherwise.} \end{cases}$$

These quantities are purely multiplicative valuations that satisfy:

$$p_k^{(\infty)}(\omega) = p_{k-1}^{(\infty)}(\omega/0)\, p_{k-1}^{(\infty)}(\omega/1). \tag{4.33}$$

Furthermore, the *expectation* $p_{k,n}^{(\infty)}$ of $p_k^{(\infty)}(\omega)$ over all ω of cardinality n is exactly the *probability* for an n-set of strings to have an associated trie of heigth at most k.

Thus these probabilities, under the Bernoulli model, have generating functions that satisfy:

$$p_k^{(\infty)}(x) = p_{k-1}^{(\infty)}\left(\frac{x}{2}\right)^2 \tag{4.34}$$

$$= p_0^{(\infty)}(x\, 2^{-k})^{2^k} = e_b(x\, 2^{-k})^{2^k}.$$

With the general relations between the Bernoulli and Poisson models that have been given in Lemma 7, we thus find the values of these probabilities under the Poisson model to be:

$$p_k^P(n) = e^{-n}\, e_b(n\, 2^{-k})^{2^k}. \tag{4.35}$$

From (4.34), (4.35), we have access to the expected height of b-tries, and we find:

Theorem F: *The average depth of a b-trie is under the Bernoulli model:*

$$de_n^{(\infty)} = \sum_{k \geq 0} (1 - n![x^n]\, e_b(x\, 2^{-k})^{2^k}),$$

and under the Poisson model of parameter n:

$$de_n^P = \sum_{k \geq 0} (1 - e^{-n}\, e_b(n\, 2^{-k})^{2^k}).$$

The asymptotic analysis of the results of Theorems E,F reveals that the expected number of pages fluctuates around

$$\frac{n}{b.\ \log 2},$$

which corresponds to a load factor of the pages close to log2 [11, 12, 2]. The depth of a b-trie under either the Bernoulli or the Poisson model satisfies:

$$de_n = (1 + \frac{1}{b}) \log_2 n + O(1).$$

In Extendible Hashing, the trie is embedded in a perfect tree, and represented as an array; the size of this array is exactly 2 raised to a power which is the height of the trie. It has been analysed under both models by [8, 16] who show that it has a *non-linear growth* and fluctuates around:

$$\frac{[(b+1)!]^{\frac{1}{b}}}{\log 2} \Gamma(1 - \frac{1}{b}) n^{1 + \frac{1}{b}}.$$

5. Miscellaneous Applications

5.1. Multidimensional Search

Multidimensional tries or $k{-}d$-tries (in dimension k) are used to store and retrieve records from a k-dimensional universe. Throughout this section, we assume that each field is an infinitely long binary string. The universe of records is then simply:

$$(\mathbf{B}^{(\infty)})^k$$

There is a very natural mapping from $(\mathbf{B}^{(\infty)})^k$ to $\mathbf{B}^{(\infty)}$. To an element $\vec{u} \subset (\mathbf{B}^{(\infty)})^k$, $\vec{u} = (u^{[1]}, u^{[2]}, \cdots ; u^{[k]})$, we associate the string:

$$v = sh(\vec{u}) = u_1{}^{[1]}u_1{}^{[2]} \ldots u_1{}^{[k]}u_2{}^{[1]}u_2{}^{[2]} \ldots$$
$$u_2{}^{[k]}u_3{}^{[1]}u_3{}^{[2]} \ldots u_3{}^{[k]} \ldots .$$

In other words, v is obtained from $\vec{u}$ by regularly shuffling the bits of the components of $\vec{u}$.

Given a finite set $\omega \in (\mathbf{B}^{(\infty)})^k$, the trie built on $sh(\omega)$ is by definition the k–d-trie associated to ω.

The use of k–d-tries should be clear. To retrieve a record when all its fields are known, follow a path in the tree guided by the bits of fields in a manner consistent with the definition of the shuffle function.

The interest of k–d-tries is to allow partial match retrieval to be performed often with reasonable efficiency. To retrieve a record with *some* of its fields specified, again follow a path in the tree guided by the bits specified in the query. For bits corresponding to unspecified fields, proceed with a search in both subtrees. This method is described in [18]. A *partial match search* is thus a succession of one-way and two-way branching dependent upon the *specification pattern* of the query.

Definition: *A specification pattern is a word of length k over the alphabet* $\{S, *\}$. *To any partial match query there corresponds a unique specification pattern obtained by associating an "S" to a specified field, and a "*" to an unspecified field in the query.*

Our purpose here is only to illustrate by means of an example extracted from [5] how previously discussed methods may be used to analyse the cost of partial match queries in k–d-tries. One has:

Theorem G: *The expected cost, measured in the number of internal nodes traversed, of a partial match query on a file of size n, represented as a k–d-trie with specification pattern π is:*

$$c_{\pi,n} = \sum_{l=0}^{k-1} [\delta_1 \delta_2 \cdots \delta_l \sum_{j \geq 0} 2^{j(k-s)} \tau_{j,l}(n)]$$

where $\delta_j = 1$ if $\pi_j = $ "S", $\delta_j = 2$ if $\pi_j = $ "*", and:

$$\tau_{j,l}(n) = 1 - (1 - \frac{1}{2^{kj+l}})^n - \frac{n}{2^{kj+l}} (1 - \frac{1}{2^{kj+l}})^{n-1},$$

for j, l not both 0; $\tau_{0,0}(n) = 0$.

Proof: Let $c_{\pi}(\omega)$ denote the expected cost of a *random* search (*i.e.* specified fields according to π in the query are drawn uniformly) in the *fixed* tree *trie* (ω). Letting $\pi^{<1>}, \pi^{<2>}, \cdots$ denote the successive left circular shifts of the word π; in particular $\pi^{<k>} = \pi^{<0>} = \pi$, $\pi^{<k+1>} \geq \pi^{<1>}, \cdots$. From the structure of the recursive search procedure, we find the recurrence:

$$c_{\pi^{<j>}}(\omega) = \frac{\delta_j}{2} c_{\pi^{<j+1>}}(\omega/0) + \frac{\delta_j}{2} c_{\pi^{<j+1>}}(\omega/1) +$$
$$+1 - \delta_{|\omega|,0} - \delta_{|\omega|,1}, \qquad (5.1)$$

which is a direct reflection of the cyclical changes of the discriminating fields in a k–d-trie.

Equations (5.1) translate into a set of difference equations for corresponding generating functions:

$$c_{\pi^{<j>}}(x) = \delta_j c_{\pi^{<j+1>}}(\frac{x}{2}) + e^x - 1 - x, \quad j=0 \cdots k-1. \qquad (5.2)$$

The system of k equations (5.2) reduces by successive elimination of $c_{\pi^{<1>}}, c_{\pi^{<2>}}, \cdots$ with $c_{\pi^{<k>}}(x) = c_{\pi}(x)$ to:

$$c_\pi(x) = 2^{k-s} \, c_\pi \left(\frac{x}{2^k}\right) + a_\pi(x), \tag{5.3}$$

where s is the number of specified attributes in the query and $a_\pi(x)$ is a combination of exponential functions.

System (5.3) is then solved by iteration as explained in Section 2, and Theorem G is then proved. ∎

The result of the asymptotic analysis of our previous estimates is that the average cost of a partial match query with s out of k attributes specified in a file of size n is

$$O(n^{1-\frac{s}{k}}).$$

This result is to be compared to the corresponding cost of a search in a $k{-}d$-tree (the multidimensional analogue of binary search trees) which is:

$$O(n^{1-\frac{s}{k} + \vartheta(\frac{s}{k})})$$

for a non-zero function $\vartheta(\frac{s}{k})$. These analyses are presented in detail in [5]. They give support to Nievergelt's claim [13] that in many situations digital structures tend to be more efficient than comparison based structures.

5.2. Biased Tries and Polynomial Factorisation

Tries constructed from a biased distribution may be analysed by the methods of Section 3.2. Additive parameters are no difficulty, especially if one uses methods of Lemma 7 (ii). See *e.g.* [11, ex. 5.5.2.53].

A generating function of a rather surprising form occurs in the analysis of the expected height of biased tries. Defining as in Section 4.3 the quantities $p_{n,k}$ to be the probability that a simple trie built on a binary alphabet has height at most k, we readily find from Section 3.2

and decomposition (4.33) the relation on the exponential generating function of the $p_{n,k}$:

$$p_k(x) = p_{k-1}(px) \cdot p_{k-1}(qx),$$ (5.4)

which, combined to the initial condition:

$$p_0(x) = 1+x$$

shows that

$$p_k(x) = \prod_{j \geq 0} (1+p^j \, q^{k-j} \, x)^{\binom{k}{j}}.$$

Whence:

Theorem H: *The probability for a biased trie formed with n strings to have height at most k is:*

$$n! [x^n] \prod_{j \geq 0} (1+p^j \, q^{k-j} \, x)^{\binom{k}{j}}.$$

In [8], the authors use a saddle point argument to show that the corresponding expected height is of order:

$$\frac{2\log n}{\log (p^2 + q^2)^{-1}}.$$

This result is useful in the analysis of some refinements of Berlekamp's polynomial factorisation algorithms based on the construction of idempotents.

6. Conclusions

It should be clear by now that a large number of statistics on binary sequences can be analysed rather simply by the methods which we have previously developed. Other applications that we do not have space to describe here include the Probabilistic Counting algorithm of

[4] or the analysis by [6] of the von Neumann-Knuth-Yao algorithm for generating an exponentially distributed variate.

More generally, consideration of general relationships between structural definitions of algorithms or combinatorial parameters and the corresponding equations satisfied by generating functions seems worthy of attention in the field of analysis of algorithms. It extends the approach of some recent works in combinatorial analysis by Foata and Schutzenberger, Rota, Jackson and Goulden [10]. That it is not restricted to tries will only be illustrated by means of a few simple examples.

Assume two valuations on binary trees are related by:

$$v(T) = w(left(T)).$$

We can then set up various translation lemmas for corresponding generating functions of average values under several statistical models of tree formation. We cite:

(i): in the case of binary tries (as we have been considering), for exponential generating functions:

$$v(z) = e^{z/2} w(\frac{z}{2}) \; ;$$

(ii): in the case of binary search trees, for ordinary generating functions:

$$v(z) = \int_0^z w(t) \frac{dt}{1-t} \; ;$$

(iii): in the case of digital trees [Kn73], for exponential generating functions:

$$v(z) = \int_0^z e^{t/2} w(\frac{t}{2}) \, dt \; ;$$

(iv): in the case of planar binary trees, for generating functions of cumulated values:

$$v(z) = \frac{1 - \sqrt{1 - 4z}}{2} \, w(z).$$

Thus, for entire classes of valuations we can characterise the systems of functional equations that arise. This characterisation calls for:

-exact resolution methods; this is provided by iteration mechanism in case (i), by the resolution of differential systems in (ii) and by the resolution of algebraic systems in case (iv);

-methods for pulling out (if possible directly from equations satisfied by generating functions) the asymptotic behaviour of coefficients; the available tools are: (i) Mellin transform techniques, (ii) contour integration in conjunction with the study of singular differential systems, (iii) Newton series and corresponding integral formulae, (iv) the Darboux method for relating singularities of functions to asymptotics of their Taylor coefficients.

References

[1] A. V. Aho, J. E. Hopcroft, and J. D. Ullman, *Data Structures and Algorithms,* Addison-Wesley, Reading, Mass. (1983).

[2] R. Fagin, J. Nievergelt, N. Pippenger, and R. Strong, *"Extendible Hashing: A Fast Access Method for Dynamic Files,"* A.C.M. T.O.D.S. 4 pp. 315-344 (1979).

[3] G. Fayolle, P. Flajolet, and M. Hofri, *"On a Functional Equation Arising in the Analysis of a Protocol for a Multiaccess Broadcast Channel,"* INRIA Res. Rep. 131 (April 1982).

[4] P. Flajolet and P. N. Martin, *"Probabilistic Counting,"* Proc. 24th I.E.E.E. Symp. on F.O.C.S., (1983).

[5] P. Flajolet and C. Puech, *"Tree Structures for Partial Match Retrieval,"* Proc. 24th I.E.E.E. Symp. on F.O.C.S., (November 1983).

[6] P. Flajolet and N. Saheb, *"Digital Search Trees and the Complexity of Generating an Exponentially Distributed Variate,"* in Proc. Coll. on Trees in Algebra and Programming, Lecture Notes in Comp. Sc., L'Aquila (1983). to appear.

[7] P. Flajolet and D. Sotteau, *"A Recursive Partitioning Process of Computer Science,"* Proc. II World Conference on Mathematics at the service of Man, pp. 25-30 (1982).

invited lecture.

[8] P. Flajolet and J-M. Steyaert, *"A Branching Process Arising in Dynamic Hashing, Trie Searching and Polynomial Factorization,"* Proceeding ICALP 82, Lect. Notes in Comp. Sc. 140, pp. 239-251 (1982).

[9] E. Fredkin, *"Trie Memory,"* Comm. ACM 3 pp. 490-499 (1960).

[10] I.P. Goulden and D. M. Jackson, *Combinatorial Enumeration,* John Wiley, New York (1983).

[11] D. E. Knuth, *The Art of Computer Programming: Sorting and Searching,* Addison-Wesley, Reading, Mass. (1973).

[12] P. A. Larson, *"Dynamic Hashing,"* BIT 18 pp. 184-201 (1978).

[13] J. Nievergelt, *"Trees as Data and File Structures,"* Proceedings CAAP 81, Lect. Notes in Comp. Sc. 112, pp. 35-45 (1981).

[14] J. Nievergelt, H. Hinterberger, and K. Sevcik, *"The Grid File: an Adaptable Symmetric Multi-key File,"* Report 46, E.T.H. (1981).

[15] L. Trabb Pardo, *"Set Representation and Set Intersection,"* Report STANCS-78-681, Stanford University (1978).

[16] M. Regnier, *"Evaluation des performances du hachage dynamique,"* Thesis, Universite Paris-Sud (1983).

[17] E. Reingold, J. Nievergelt, and N. Deo, *Combinatorial Algorithms,* Prentice-Hall, Englewood Cliffs, N. J. (1977).

[18] R. L. Rivest, *"Partial Match Retrieval Algorithms,"* S.I.A.M. J. on Comp. 5 pp. 19-50 (1976).

[19] R. Sedgewick, *Algorithms,* Addison-Wesley, Reading, Mass. (1983).

[20] T. A. Standish, *Data Structure Techniques,* Addison-Wesley, Reading, Mass. (1980).

[21] E. H. Sussenguth, *"Use of Tree Structures for Processing Files,"* Comm. A.C.M. 6(5) pp. 272-279 (1963).

Annals of Discrete Mathematics 25 (1985) 189-210
© Elsevier Science Publishers B.V. (North-Holland)

EASY SOLUTIONS FOR THE K–CENTER PROBLEM OR THE DOMINATING SET PROBLEM ON RANDOM GRAPHS

Dorit S. HOCHBAUM

School of Business Administration
University of California, Berkeley

This paper is concerned with the probabilistic behavior of three NP–complete problems: the K-center, the dominating set, and the set covering problem. There are two main contributions: one is the asymptotic values of the optimal solutions and the probability expressions for the interval of values where the optimal solution value lies for every finite problem size. The other contribution is the proof that *any* random feasible solution is asymptotically optimal with probability that is asymptotically 1.

§ 1. Introduction

This paper describes an analysis of the optimal solution and approximate solution values of the K-center, dominating set, and set-covering problems. All these problems are known to be NP–complete (see Garey and Johnson [4] for definitions). Moreover, finding an approximate solution to the K-center problem within fixed error ratio is itself an NP–complete problem [5], and a similar result is likely to apply to the dominating set and set-covering problem, although this question is still open. Though the concept of NP–completeness indicates that it is most likely that these problems are inherently hard to solve, this paper proves that in certain random environments these problems almost surely become "ridiculously easy" to solve asymptotically. In fact, the solution amounts to picking indiscriminately a feasible solution. Another helpful result in this paper is the asymptotic value of the optimal solution as well as the range of likely values for finite instances.

Such a surprising result about the ease of solving the problems asymptotically using any feasible solution is rather rare. It has been reported by Burkard and Fincke [1, 2] for the quadratic bottleneck and sum assignment problems and other problems with similar properties. Another similar result by Vercellis [7] will be discussed in more detail in the section describing the set-covering results.

It should be noted that such results imply that empirical tests based on random problems drawn from the distributions analyzed are not necessarily very meaningful (in terms of proving the quality of certain algorithms) unless they provide evidence of superior performance compared to the trivial selection algorithm analyzed here.

We now define the three problems analyzed in this paper. The K-center problem appears frequently in the context of locating emergency facilities. Given n locations with distance (or other forms of cost function) associated with each pair, the K-center problem is to find a subset of size K of the set locations such that the maximum distance between any location and the "nearest" facility is minimized. For instance, when the facilities are hospitals, we do not want any potential patient to be too far from the nearest hospital.

The dominating set problem is defined on a graph $G = (V,E)$. The problem is to identify the smallest possible subset of vertices S with the property that for every vertex v in $V - S$ there exists some edge in E connecting v to a vertex of S. A subset S with this property is called a dominating set. In the format of a decision problem, the dominating set is: given a value K, is there a dominating set of size K or less?

The set-covering problem is a generalization of the dominating set problem. Given any set of elements to be covered and a collection of subsets $S_1 \ldots S_m$, find a subcollection of minimum number of subsets such that every element belongs to at least one of the subsets in the subcollection.

The main theorems are presented for the dominating set problem. The generalization to the set-covering problem is fairly straightforward, and the corresponding theorem for the K-center problem follows from the following reduction.

The Reduction

Every instance of the K-center problem is reducible to an instance of the dominating set problem. As a decision problem the K-center problem is whether there is a subset of vertices of size K, such that the maximum weight on the edge between each vertex and the nearest vertex in the subset is lesser or equal to M. Eliminating all edges the weights of which exceed M, creates a graph that has a dominating set of size K, if and only if the answer to the K-center decision problem is affirmative.

This reduction is conveniently represented for the random structures analyzed. We denote the density function by f and the cumulative distribution function for the edge weights by F. For each random occurrence of G_n (f) the K-center decision problem is transformed to the dominating set decision problem on a random occurrence of $G_{n,p}$ with $p = F(M)$. Solving the K-center problem as a minimization problem on G_n (f) amounts to finding a minimum value of p such that there is a K-dominating set in the corresponding instance of $G_{n,p}$. The corresponding minimum value of the K-center problem is then F^{-1} (p).

Note that this reduction is reversible, i.e., solving the dominating set problem on $G_{n,p}$ amounts to finding a minimum value of K such that the solution value or the K-center problem does not exceed $F^{-1}(p)$ on a graph with weights lesser than $F^{-1}(p)$ assigned to all edges that exist in the graph $G_{n,p}$ and larger than $F^{-1}(p)$ assigned to all other edges.

The Random Structures

The K-center problem can be viewed as defined on a complete graph with n vertices representing locations and edge weights representing the distances between all pairs. We assume that the edge weights associated with each pair of locations are independently and identically distributed (i.i.d.) random variables. Furthermore, we assume that these random variables are distributed according to a uniformly positive density function f defined on an interval, (i.e., $\exists\, \varepsilon > 0$ such that $f(\omega) > \varepsilon$ for all values ω in the sample space). Such density functions include the uniform density, truncated normal density and almost all common density functions with support in a compact interval.

The K-center problem is analyzed for large n. We study a sequence of graphs parameterized by n and f — G_n (f).

For each value of n we assume that the weights on the edges of the complete graph, on n vertices, are regenerated according to the prescribed distribution. Such a model is called the independent model as opposed to the incremental model, where for each larger graph only the *additional* edge weights are generated. This distinction, pointed out by Weide [8], suggests that our results are strong as compared to similar analyses of the incremental model.

The dominating set problem is analyzed for a family of *p-random graphs* $G_{n,p} = (V_n, E_{n,p})$. A p-random graph $G_{n,p}$ is a graph on n vertices where each edge belongs to $E_{n,p}$ with probability p. We assume that the graphs $G_{n,p}$ are generated, like the family of graphs $G_n(f)$, according to the independent model.

Overview of Results

In section 2 we analyze the dominating set problem as a decision problem on p-random graphs. Here the technical result is given in the lemma that establishes a narrow range for the values of the probability of existence of a K-dominating set in $G_{n,p}$. We find, in theorem 1, threshold values for p, $p_U(n,K)$ and $p_L(n,K)$ such that whenever $p > p_U(n,K)$ there is a K-dominating set in a p-random graph $G_{n,p}$ almost surely, and whenever $p < p_L (n,K)$ there is no K-dominating set in $G_{n,p}$. It is also proved that for any fixed K the asymptotic behavior of $p_U (n,K)$ and $p_L (n,K)$ is identical to that of $1 - 1/(n-K)^{1/K}$.

As a corollary of theorem 1 the following statements are proved to hold almost surely:* There is no dominating set of size K when $K = o(\ell n\ n)$; there is a dominating set of size K when $K = \Omega(\ell n\ n)$; when $K = c \cdot \ell n\ n$ there is a dominating set of size K if $p > p_U(n, c \cdot \ell n\ n)$, where $\lim_{n \to \infty} p_U(n, c \cdot \ell n\ n) = 1 - e^{-1/c}$, and there is no dominating set of size K if $p < p_L(n, c \cdot \ell n)$, where $\lim_{n \to \infty} p_L(n, c \cdot \ell n\ n) = 1 - e^{-1/c}$, (i.e., this is a threshold type result).

The relative ease with which we can solve the dominating set problem derives from the result stated in theorem 2; that *any* random K-subset pf vertices in $G_{n,p}$ is a dominating set with probability $1 - 1/n^\gamma$, for $p > \bar{p}(n, K, \gamma)$. We also prove that $\bar{p}(n, K, \gamma)$ is $0(1 - 1/[n^\gamma \cdot (n\text{-}K)]^{1/K})$. This statement implies that the asymptotic behavior of $\bar{p}(n, K, \gamma)$ is similar to that of $p_U(n, K)$. Hence, given a value of p that exceeds $1 - e^{-(1 + \gamma)/c}$, the dominating set problem is solved by a random selection algorithm with probability $1 - 1/n^\gamma$.

Section 3 describes the implications of theorems 1 and 2 to the dominating set as an optimization problem. In this section we show that the asymptotic size of the minimum dominating set, in a p-random graph, is $\ell n\ n / \ell n(\frac{1}{1\text{-}p})$ almost surely. We consider now the random selection algorithm, i.e., an algorithm that picks vertices randomly until these vertices constitute a dominating set. The performance of the random selection algorithm on a p-random graph $G_{n,p}$ is established in theorem 3—— it yields a solution with relative error less than ε with probability $1 - 1/n^\varepsilon$. We extend this result to the set-covering problem and show that the same result holds for the random selection algorithm that finds a solution to the set-covering problem. This extension is outlined in section 4.

---

* We use the notation o, 0, and Ω to denote for real functions f(x) and g(x):

(1) f is said to be o(g) iff $\lim_{x \to \infty} f(x)/g(x) = 0$

(2) f is said to be 0(g) iff $\lim_{x \to \infty} f(x)/g(x) = constant$

(3) f is said to be Ω (g) iff $\lim_{x \to \infty} g(x)/f(x) = 0$

In section 5 we use the reduction of the K-center problem to the dominating set problem. Here we analyze a random selection algorithm that picks randomly any K-subset. Writing $K = c \cdot \ln n$ we prove that the error, due to the random solution as compared to the optimum, is $O(e^{-1/c}(1-e^{-\gamma/c}))$ with probability $1 - 1/n^{\gamma}$.

The conclusion is that the error expression is arbitrarily close to zero almost surely for c sufficiently large or sufficiently small, that is whenever K is not proportional to log n (the coefficient c for which the absolute error is maximum can be computed precisely for each given γ). Note that this case is a relatively less interesting set of instances of the K-center problem since it can be solved using enumeration in subexponential time.

§2. Analysis of the dominating set decision problem

We consider the dominating set problem on a p-random graph on n vertices, $G_{n,p}$. The following lemma is the backbone of all the results derived in this paper.

Lemma 1

Let the probability that a p-random graph on n vertices contains a K dominating set be denoted by $p_K(n, p)$. Then

$$1 -\left\{1 -[1-(1-p)^{K-1}]^{n-K}\right\}^{\lfloor \frac{n}{K} \rfloor} \leqslant P_K(n,p) \leqslant$$
$$1- \left\{1-[1-(1-p)^{K}]^{n-K}\right\}^{\binom{n}{K}}$$

Proof

First, we prove the upper bound inequality. Given an arbitrary subset of size K, S, the probability that there is at least one edge from a vertex not in S to some vertex in S is $1 - (1-p)^K$. The probability that there is at least one edge between every vertex in V-S and some vertex in S is $[1 - (1-p)^K]^{n-K}$, i.e., S is dominating. Let P(S n.d.) denote the probability that S is a nondominating set. We want to evaluate

$$1 - P(S_1 \text{ n.d.} \wedge S_2 \text{ n.d.} \wedge \ldots \wedge S_{\binom{n}{K}} \text{ n.d.})$$

for all possible subsets of size K.

$$P(S_1 \text{ n.d.} \wedge S_2 \text{ n.d.} \wedge \ldots \wedge S_{\binom{n}{K}} \text{ n.d.})$$

$$= P(S_1 \text{ n.d.}) \cdot P(S_2 \text{ n.d.}/S_1 \text{ n.d.}) \cdot \ldots$$

$$\cdot P(S_{\binom{n}{K}} \text{ n.d.}/S_1 \text{ n.d.} \wedge S_2 \text{ n.d.} \wedge \ldots \wedge S_{\binom{n}{K}-1} \text{ n.d.})$$

Given that there is a nondominating set in a graph, the likelihood of another set being nondominating is either unaffected or could only increase, as compared to a random graph where the existence of a nondominating set is not known. This fact is proved precisely in Claim 2. Therefore,

$$P(S_1 \text{ n.d.} \wedge \ldots \wedge S_{\binom{n}{K}} \text{ n.d.}) \geq P(S_1 \text{ n.d.}) \cdot$$

$$\cdot P(S_2 \text{ n.d.}) \cdot \ldots P(S_{\binom{n}{K}} \text{ n.d.})$$

$$= \left\{ 1 - [(1-p)^K]^{n-K} \right\}^{\binom{n}{K}}$$

Hence, the upper bound inequality is established.

Before proceeding to the proof of the lower bound, we shall define the property of nondominance and introduce some notations. A set T is called nondominating on a set S (denoted by T n.d. S), if in the bipartite graph induced on the vertices of T and S-T respectively there exists a vertex in S-T with no neighbor in T. Such vertex will be henceforth referred to as *free*, and a set that contains only free (nonfree) vertices will be called a *free* (nonfree) *set*. A set will simply be called n.d. if it does not dominate V. Let $S_1, S_2, \ldots, S_{\lfloor \frac{n}{K} \rfloor}$ be any collection of disjoint sets of size K each.

Claim 1

Let S_{j_1}, S_{j_2} be two disjoint sets of size K each, then

$$Pr(S_{j_1} \text{ n.d. } S_{j_2} \mid S_{j_2} \text{ n.d.}) \leqslant 1 - [1 - (1 - P)^{K-1}]^K.$$

Proof

The proof will make use of the following notations:

$E_i = Pr$ (exactly i vertices in S_{j_1} are free), i = 0, 1, ..., K

$P_i = Pr$ (S_{j_1} n.d. S_{j_2}/exactly i vertices in S_{j_1} are free), i = 0,1,...,K

$E^u = Pr$ (at least u is free), $u \in S_{j_1}$

$P^u = Pr$ (S_{j_1} n.d. S_{j_2}/at least u is free), $u \in S_{j_1}$.

Let u be one specific vertex in S_{j_1}.

$$1 - [1-(1-p)^{K-1}]^K = P^u = \frac{KP^u \cdot E^u}{K \cdot E^u} = \frac{\sum_{u \in S_{j_1}} P^u E^u}{\sum_{u \in S_{j_1}} E^u}$$

$$= \frac{\sum_{i=0}^{K} iE_i P_i}{\sum_{i=0}^{K} iE_i} \overset{(*)}{\geqslant} \frac{\sum_{i=0}^{K} E_i P_i}{\sum_{i=0}^{K} E_i} = Pr(S_{j_1} \text{ n.d. } S_{j_2}/S_{j_2} \text{ n.d.}).$$

The inequality (*) follows from claim 2 below establishing that $P_0 \leqslant P_1 \leqslant P_2 \leqslant \ldots \leqslant P_{K-1} \leqslant P_K$; multiplying the inequality (*) by the denominators, it is equivalent to $\displaystyle\sum_{i,j=0}^{K} (P_i - P_j) \cdot (i-j) \geqslant 0$. This

inequality holds because of the monotonicity of P_i.

Claim 2

$$P_0 \leqslant P_1 \leqslant P_2 \leqslant \ldots \leqslant P_K$$

Proof

We evaluate P_i for a specific set of i vertices. Due to symmetry reasons this is also the probability expression for every set of i vertices. Let $T \cup \{v\} \subseteq S_{j_1}$ and $|T| = i$.

$1 - P_{K-i-1} = \Pr(S_{j_1}$ dominates $S_{j_2}/$exactly $T \cup \{v\}$ nonfree$)$

$$= \frac{\Pr(S_{j_1} \text{ dominates } S_{j_2} \cap \text{ exactly } T \cup \{v\} \text{ nonfree})}{\Pr(T \text{ nonfree} \cap S_{j_1}\text{-}T\text{-}\{v\} \text{ free}) \cdot \Pr(v \text{ nonfree})}.$$

The event $\left\{T \cup \{v\} \text{ dominates } S_{j_2} \text{ and } T \cup \{v\} \text{ nonfree}\right\}$ contains the event $\left\{T \text{ dominates } S_{j_2} \cap T \text{ nonfree and } S_{j_1}\text{-}T\text{-}\{v\} \text{ free} \cap v \text{ nonfree}\right\}$.

Hence,

$1 - P_{K-i-1} \geqslant$

$$\geqslant \frac{\Pr(T \text{ dominates } S_{j_2} \cap T \text{ nonfree and } S_{j_1}\text{-}T\text{-}\{v\} \text{ free}) \cdot \Pr(v \text{ nonfree})}{\Pr(T \text{ nonfree and } S_{j_1}\text{-}T\text{-}\{v\} \text{ free}) \cdot \Pr(v \text{ nonfree})}$$

$$= \frac{\Pr(T \text{ dominates } S_{j_2} \cap T \text{ nonfree and } S_{j_1}\text{-}T\text{-}\{v\} \text{ free}) \cdot \Pr(v \text{ free})}{\Pr(T \text{ nonfree and } S_{j_1}\text{-}T\text{-}\{v\} \text{ free}) \cdot \Pr(v \text{ free})}$$

$$= \frac{\Pr(T \text{ dominates } S_{j_2} \cap T \text{ nonfree})}{\Pr(T \text{ nonfree})} =$$

$$= \Pr(Sj_1 \text{ dominates } S_{j_2}/\text{exactly } S_{j_1}\text{-T free})$$

$$= 1 - P_{K-i} \cdot$$

Therefore, $P_{K-i} \geqslant P_{K-i-1}$ for $i = 0, \ldots, K\text{-}1$ as claimed.

Q.E.D.

We now evaluate the probability for S_j to be nondominating conditioned on $S_1, \ldots, S_{j-1}$ being nondominating.

$$\Pr(S_j \text{ n.d.}/S_1 \text{ n.d.}, S_2 \text{ n.d.}, \ldots, S_{j-1} \text{ n.d.})$$

$$= \Pr\big((S_j \text{ n.d. on } V - \{S_1 \cup S_2 \cup \ldots \cup S_{j-1}\}) \cup (S_j \text{ n.d. on } S_1) \cup \ldots$$

$$\ldots \cup (S_j \text{ n.d. on } S_{j-1})/S_1 \text{ n.d.}, S_2 \text{ n.d.}, \ldots, S_{j-1} \text{ n.d.}\big)$$

$$= \Pr\big((S_j \text{ n.d. on } V - \{S_1 \cup S_2 \cup \ldots \cup S_{j-1}\}) \cup (S_j \text{ n.d. } S_1/S_1 \text{ n.d.}) \cup \ldots$$

$$\ldots \cup (S_j \text{ n.d. } S_{j-1}/S_{j-1} \text{ n.d.})\big).$$

Since the events of S_j being nondominating on disjoint subset of V are independent, the above is equal to

$$= 1 - \big(1 - \Pr(S_j \text{ n.d. on } V - \{S_1 \cup S_2 \cup \ldots \cup S_{j-1}\})\big) \cdot$$

$$\cdot \big(1 - \Pr(S_j \text{ n.d. on } S_1/S_1 \text{ n.d.})\big) \cdot \big(1 - \Pr(S_j \text{ n.d. on } S_2/S_2 \text{ n.d.})\big) \cdot \ldots$$

$$\ldots \cdot \big(1 - \Pr(S_j \text{ n.d. on } S_{j-1}/S_{j-1} \text{ n.d.})\big)$$

$$\leq 1 - [1 - (1-p)^K]^{n-jK} \cdot [1-(1-p)^{K-1}]^K \ldots [1-(1-p)^{K-1}]^K$$

$$= 1 - [1-(1-p)^K]^{n-jK} \cdot [1-(1-p)^{K-1}]^{(j-1)K}$$

$$\leq 1 - [1-(1-p)^{K-1}]^{n-K}$$

Finally, these inequalities are applied to obtain the lower bound expression:

$$\Pr(S_1 \text{ n.d.}, S_2 \text{ n.d.}, \ldots, S_{\lfloor\frac{n}{K}\rfloor} \text{ n.d.})$$

$$= \Pr(S_1 \text{ n.d.}) \cdot \Pr(S_2 \text{ n.d.}/S_1 \text{ n.d.}) \cdot \ldots$$

$$\cdot \Pr(S_{\lfloor\frac{n}{K}\rfloor} \text{ n.d.}/S_1 \text{ n.d.}, \ldots S_{\lfloor\frac{n}{K}\rfloor} \text{ n.d.})$$

$$\leq \left\{ 1-[1-(1-p)^{K-1}]^{n-K} \right\}^{\lceil\frac{n}{K}\rceil}$$

Q.E.D.

The probability expressions derived in this lemma may be used to establish the range of values (confidence interval) for the size of the minimum dominating set. The lower end of such interval is determined when the upper bound probability expression is close to zero, and the upper end when the lower bound probability expression is close to 1. For instance, in the case of $n = 1000$, $p = 1/2$, the minimum dominating set could assume one of only five values with probability exceeding 0.9. It follows from theorem 1 below that this range of values shrinks asymptotically with n.

Theorem 1

(a)   There is, almost surely, a K dominating set in $G_{n,p}$, if $p \geq p_U(n,K)$

(b)   There is, almost surely, no K dominating set in $G_{n,p}$, if $p \leq p_L(n,K)$

(c) $p_U(n,K) = 0(1 - 1/(n - K)^{1/K})$ and

$$p_L \, (n, K) = 0(1 - 1/(n - K)^{1/(K-1)}).$$

Proof

(a) We use the inequalities provided by Lemma 1 to find a minimal value of p that guarantees that both the expression on the left and on the right exceed $1 - \varepsilon$ where ε is $\dfrac{1}{n^\gamma}$.

In order for the expression on the right to exceed $1 - \varepsilon$, p has to satisfy $p > 1 - [1 - (1 - \varepsilon^{1/\binom{n}{K}})^{1/(n-K)}]^{1/K}$.

Now, $\varepsilon^{1/\binom{n}{K}} = e^{-\gamma \, \ell n \, n/\binom{n}{K}} \cong 1 - \dfrac{\gamma \, \ell n \, n}{\binom{n}{K}}$. Substituting in the lower bound expression above for p, we have

$$p \geq 1 - [1 - (\frac{\gamma \, \ell n \, n}{\binom{n}{K}})^{1/(n-K)}]^{1/K} \ .$$

We rewrite $(\dfrac{\gamma \, \ell n \, n}{\binom{n}{K}})^{1/(n-K)}$ as $e^{-\frac{1}{n-K} (\ell n \, \binom{n}{K} - \ell n \, \ell n \, n - \ell n \, \gamma)}$

which can be approximated by $1 - \dfrac{1}{n-k} (\ell n \binom{n}{K} - \ell n \, \ell n \, n - \ell n \, \gamma)$.

Hence,

$$[1 - (\frac{\gamma \, \ell n \, n}{\binom{n}{K}})^{1/n-K}]^{1/K} \cong [\frac{\ell n \binom{n}{K} - \ell n \, \ell n \, n \, \ell n \, \gamma}{n - K}]^{1/K} \quad \text{and}$$

$$p \geq 1 - (\frac{\ell n \binom{n}{K}}{n-K})^{1/K}$$ will satisfy the convergence of the right-hand side

to 1 as $1 - 1/n^\gamma$ for any fixed γ. When $K = \Omega(\ell n \, n)$ any fixed positive value of $p > 0$ will asymptotically exceed the lower bound. More precisely, for any value of n, p exceeding $1 - e^{\dfrac{-\ell n(n-K)}{K}} \cong \dfrac{\ell n(n-K)}{K}$ guarantees the convergence of the right-hand side to 1 at a rate equal to or faster than $1 - \varepsilon$. When $K = c \cdot \ell n\ n$ for some $c > 0$ the lower bound for p converges to $1 - e^{-1/c}$. When $K = o(\ell n\ n); p > e^{-\ell n(n-k)/K}$; therefore, only very dense graphs will have a K dominating set in the latter case.

Following similar arguments for the expression on the left-hand side yields

$$p \geq 1 - [1 - (1 - e^{\dfrac{-\gamma\ K\ \ell n\ n}{n}})^{1/n-K}]^{1/(K-1)}.$$

(Note that without loss of generality $\left\lfloor \dfrac{n}{K} \right\rfloor$ may be replaced by $\dfrac{n}{K}$ as the asymptotic form of the expression is not affected.) We follow a straightforward analysis of the expression above for three cases: when $K \cdot \ell n\ n = \Omega(n)$; when $K \cdot \ell n\ n = 0\ (n)$ and when $K \cdot \ell n\ n = o(n)$. As a result, we determine that it suffices for p to exceed $1 - e^{\dfrac{-\ell n(n-K)}{K-1}}$ in order for the probability on the left-hand side to exceed $1 - 1/n^{\gamma}$ for large values of n. Therefore, there exists a K-dominating set in a p-random graph almost surely for $p \geq 1 - \dfrac{1}{(n-K)^{1/(K-1)}}$.

(b) The condition of the nonexistence of a dominating set of size K is similarly derived by setting $1 - \left\{1 - [1-(1-p)^K]^{n-K}\right\}^{\binom{n}{K}} < \varepsilon$ and solving for p asymptotically. Such an exercise yields $p < 1 -$

$- [\dfrac{\ell n(\frac{n}{K}) + \ell n\ \frac{1}{\varepsilon}}{n-K}]^{1/K}$; this goes to zero whenever $K = o(\ell n\ n)$, to $1 - e^{-1/c}$ whenever $K = c \cdot \ell n\ n$ and to 1 whenever $K = \Omega(\ell n\ n)$.

(c) Follows from (a) and (b).

Q.E.D.

Corollary 1

For any fixed value of $0 < p < 1$ and a p-random graph on n vertices:

(1)   There is no dominating set of size K for $K = o(\ell n\ n)$ almost surely.

(2)   There is a dominating set of size K for $K = \Omega(\ell n\ n)$ almost surely.

(3)   Whenever $K = c \cdot \ell n\ n$ there is a dominating set of size K almost surely whenever $p \geq p_U(n, c \cdot \ell n\ n)$. Furthermore,

$$\lim_{n\to\infty} p_U(n, c \cdot \ell n\ n) = \lim_{n\to\infty} p_L(n,c \cdot \ell n\ n) = 1 - e^{-1/c}.$$

Proof

(1) $\dfrac{1}{(n\text{-}K)^{1/(K\text{-}1)}}$ is $0\ (\dfrac{1}{n^{1/(K\text{-}1)}})$, and for $K = o(\ell n\ n)$ this function

goes to zero with n. Hence p has to exceed a value that is asymptotically close to 1 in order for a dominating set of size K to exist.

(2) $\dfrac{1}{(n\text{-}K)^{1/(K\text{-}1)}} = e^{-\frac{\ell n(n\text{-}K)}{(K\text{-}1)}} \cong 1 - \dfrac{\ell n(n\text{-}K)}{(K\text{-}1)}.$

p has therefore to exceed $\dfrac{\ell n\ (n\text{-}K)}{(K-1)}$ when $K = \Omega\ (\ell n\ n)$; in that case the function goes to zero so any fixed value of p exceeds it for sufficiently large n.

(3)   When $K = c \cdot \ell n \; n$ $\ell n(n-K)/(K-1)$ goes to $1/c$ with n.

Q.E.D.

The next theorem establishes an interval for p such that any random K subset is a dominating set in a p-random graph with probability $1 - 1/n^{\gamma}$.

Theorem 2

Given a p-random graph on n vertices. Any random K-subset of the vertices is a dominating set with probability $1 - \dfrac{1}{n^{\gamma}}$ whenever the value of p exceeds $\bar{p}(n, K, \gamma)$. In addition, $\bar{p}(n, K, \gamma)$ is $0(1 - 1/[n^{\gamma}(n-K)]^{1/K})$.

Proof

The probability that any random K-subset is a dominating set in a p-random graph is $(1 - (1-p)^K)^{n-K}$. We will establish a range for the value of p such that the probability exceeds $1 - \varepsilon$ for $\varepsilon = 1/n^{\gamma}$. Rearranging the inequality $(1 - (1-p)^K)^{n-K} \geq 1 - \varepsilon$ one gets: $p \geq 1 - [1 - (1 - \varepsilon)^{1/(n-K)}]^{1/K}$.
Now,

$$1 - [1 - (1 - \varepsilon)^{1/n-K}]^{1/K} \cong 1 - [1 - (\tfrac{1}{e})^{\varepsilon/(n-K)}]^{1/K} \cong 1 - (\tfrac{\varepsilon}{n-K})^{1/K}.$$

The first asymptotic equality holds using $(1 - \varepsilon)^{1/\varepsilon} \cong \dfrac{1}{e}$ for ε going to zero the second holds by the fact that $e^{-x} \cong 1 - x$ for x sufficiently small.

Substituting $1/n^{\gamma}$ for ε we obtain

$$p \geq 1 - e^{\dfrac{-\ell n(n-K) + \gamma \, \ell n \; n}{K}}$$

Note that when $K = \Omega(\ell n\ n)$ the value of p has to satisfy $p \geq$ $\dfrac{\ell n(n-K) + \gamma\ \ell n\ n}{K}$, a lower bound which goes to zero with n. That means that for any fixed positive value of p, any random K-subset is a dominating set. When $K = o(\ell n\ n)$ the lower bound for p goes to 1 with n, which indicates that a random K-subset in a p-random graph for p fixed is not likely to be a dominating set.

The case when K grows at the same rate as $\ell n\ n$ is a borderline case. If $K = c \cdot \ell n\ n$ then for $p \geq 1 - e^{-(1+\gamma)/c}$ a random K-subset is a dominating set. In the range $1 - e^{-1/c} < p < 1\ e^{-(1+\gamma)/c}$ the random choice algorithm is not likely to deliver a dominating set of size lesser or equal to K, in which case other algorithms must be employed.

§3. The minimum domintating set in a random graph

The analysis provided in the proof of theorem 1 can be used to derive the minimum value K such that there is a K-dominating set almost surely. For a fixed value of p there is a K dominating set in $G_{n,p}$ almost surely when $p \geq p_U(n, k)$. We set $p = p_U(n, k)$ and solve for K asymptotically. Hence, K is $\ell n\ n/\ell n\ (\dfrac{1}{1-p})$. We repeat the same approach for solving $p = p_L(n, k)$ with the outcome, $K = \ell n\ n/\ell n(\dfrac{1}{1-p})+1$.

Therefore there is no K dominating set of size lesser than $\left\lfloor \ell n\ n/\ell n\ (\dfrac{1}{1-p}) \right\rfloor$ almost surely, and there is one for all values of K exceeding $\left\lfloor \ell n\ n/\ell n\ (\dfrac{1}{1-p}) \right\rfloor + 1$. We have observed empirically however, that the convergence to the asymptotic value is rather slow. A better estimate for the size, of the minimum dominating set in $G_{n,p}$, for finite n, is obtained by solving for K the equation

$$K \cdot \ell n\ (\frac{1}{1-p}) = \ell n\ n - \ell n \ell n\ (\frac{n}{K}).$$

Replacing $\binom{n}{K}$ by $\frac{n}{K}$ yields a poorer estimate, even though asymptotically the two expressions are identical. This observation implies that the upper bound on $P_K(n, p)$ in the proof of theorem 1 is tighter than the lower bound. In particular, that means that the number of different dominating sets in $G_{n,p}$ is closer to $\binom{n}{K}$ than to $\left\lfloor \frac{n}{K} \right\rfloor$.

The almost sure value of the minimum dominating set holds for random connected graphs. Erdös and Rényi's·[1] result states that if $p \cong \dfrac{a \cdot \ell n\ n}{n}$ for $a > 1$, then the p-random graph $G_{n,p}$ is almost surely connceted. Thus, we have the following corollary.

Corollary 2

In a p-random connected graph the size of the minimum dominating set is $\lfloor n/a \rfloor$ or $\lfloor n/a \rfloor + 1$, almost surely, where $p = \dfrac{a \cdot \ell n\ n}{n}$.

Proof

Substituting for p in the expression $\ell n\ n / \ell n\ (\frac{1}{1-p})$ yields the value n/a.

$$Q.E.D.$$

We use similar analysis to derive the minimum size of a random dominating set. That is, we set $p = \bar{p}\ (n, K, \gamma)$, defined in theorem 2, and solve for K.

The value of the minimum dominating set randomly selected is thus either $\left\lfloor (1 + \gamma) \cdot \ell n\ n / \ell n\ (\frac{1}{1-p}) \right\rfloor$ or $\left\lfloor (1+\gamma) \cdot \ell n\ n / \ell n\ (\frac{1}{1-p}) \right\rfloor + 1$ with probability $1 - 1/n^\gamma$.

Theorem 3

The random selection algorithm yields an approximate solution to be dominating set problem on $G_{n,p}$ with relative error ε, with probability $1 - 1/n^{\varepsilon}$.

Proof

Let the random solution value be denoted DS^H and the optimal solution value DS^* and define the relative error as $\dfrac{DS^H - DS^*}{DS^*}$. Since the almost sure value of DS^* is $\ell n \ n/\ell n \ (\dfrac{1}{1-p})$, the relative error is equal to γ with probability $1 - 1/n^{\gamma}$ and lesser than or equal to ε with probability $1 - 1/n^{\varepsilon}$.

Q.E.D.

Even though theorem 3 indicates that the relative error due to the random selection algorithm is arbitrarily small with probability going to 1, the rate of convergence is very slow, enough to make such algorithms impractical.

To illustrate this statement, we note that, in order for to go to zero while the probability goes to one, ε has to behave as $\Omega \ (\dfrac{1}{\ell n \ n})$. For instance, $\varepsilon = \dfrac{1}{\sqrt{\ell n \ n}}$ satisfies this requirement. It is unfortunate, though, that for this property to be satisfied for reasonably small values of ε and sufficiently large probabilities, the value of n has to be beyond the range of any practical problem instances.

§4. The set-covering problem

The dominating set problem is a special case of the set covering problem. Consider an instance with λn columns and n rows and probability p for an entry to be equal to 1. A cover is a subset of columns such that each row is incident to at least one of the columns in the subset. Theorem 1 applies almost identically with the bounds on the probability reading

$$1 - \left\{ 1 - [1-(1-p)^{K-1}]^n \right\}^{\frac{\lambda n}{K}} \le \text{Prob} \left(\begin{array}{l} \text{there exists a} \\ \text{cover of size K} \end{array} \right)$$

$$\le 1 - \left\{ 1 - [1-(1-p)^K]^n \right\}^{(\frac{\lambda n}{K})}.$$

Following precisely the same analysis as in theorem 1 for the expression on the right hand side, for instance, we obtain

$$p \ge 1 - \left(\frac{\ell n(\frac{\lambda n}{K})}{n} \right)^{1/K}.$$

The implied size of an asymptotic optimal cover is $\ell n \; n/\ell n \; (\frac{1}{1-p})$ (rounded up or down). Note, however, that the rate of convergence is affected by the presence of $\lambda \ne 1$ and is generally slower than the corresponding rate for the dominating set solution.

Theorem 2 also applies with the probability for a random selection of columns to consist of a cover being $[1-(1-p)^K]^n$. Hence, the size of a random cover is $(1+\gamma) \ell n \; n/\ell n \; (\frac{1}{1-p})$ with probability $1 - 1/n^\gamma$. The probability that the relative error will be less than or equal to ε is therefore $1 - \frac{1}{n^\varepsilon}$. This result conflicts with Gimpel's result that claims relative error lesser than ε almost surely. (Gimpel's result is referred to by Karp in [3].)

In an independent work [7], Vercellis analyzes two ranges for the error: $0 < \varepsilon \le 1$ and $1 < \varepsilon$. He notes that for the first case the convergence is with probability going to 1, where in the second case the convergence is almost sure. This result is also evident from the form of the probability expression $1 - \frac{1}{n^\varepsilon}$ stated above. Moreover, this expression underscores the continuous change in the rate of convergence as a function of ε.

§5. The K-center problem

All the results for the dominating set problem apply to the K-center problem by using a simple transformation as follows: the K-center problem is defined on a complete graph with randomly drawn weights assigned to all edges. Answering the question whether there is a solution to the K-center problem the value of which does not exceed a constant M, amounts to finding a dominating set of size K or less in the random graph derived from the complete graph by removing all arcs the weights of which exceed M. Such a graph is a p-random graph with $p = F(M)$. The minimum value associated with the optimal solution to the K-center problem corresponds, via the cumulative distribution function F, to a minimum value of p such that there is a dominating set of size K in a p-random graph.

Recalling the assumption of "uniform positivity" of the density function f, we conclude that the derivative of the inverse of the cumulative distribution function, F^{-1}, is bounded from above by a finite number $\overline{M}$.

It is convenient to express theorem 4 in terms of values of K equal to $c \cdot \ln n$. c could be a function of n going to infinity or zero with n, or a constant independent of n. In this theorem "error" means the absolute error.

Theorem 4

(a)   For $K = c \cdot \ln n$ the error term for the random solution to the K-center problem is at most $\overline{M} \cdot e^{-1/c} (1 - e^{-\gamma/c})$ with probability $1 - 1/n^{\gamma}$.

(b)   For $\gamma = 2$ (a value sufficient to guarantee almost sure convergence) the error bound term attains its maximum when $c = 2/\ln 3$, and the maximum value is $0.389 \cdot \overline{M}$.

(c)   The error expression goes to zero whenever c is going to infinity or to zero.

Proof

(a)   The optimal value is: $F^{-1}(1 - e^{-\ln n/K}) = F^{-1}(1 - e^{-1/c})$. The random solution value is asymptotically equal to F^{-1} $(1-e^{-(1+\gamma)/c})$. Hence, the error is

$$F^{-1} (1 - e^{-(1+\gamma)/c}) - F^{-1} (1 - e^{-1/c}) \leq \bar{M} \cdot e^{-1/c} (1 - e^{-\gamma/c}).$$

(b)   Easily verifiable by finding the maximum of the error expression in (a).

(c)   Obvious.

It should be noted that relative error was not studied for the K-Center problem, the reason being that optimal value could be arbitrarily close to zero. Thus, the relative error is a meaningless measure for the quality of approximation algorithms for the K-center problem with the given type of weight distribution, when arbitrarily close to zero values are permitted.

References

[1]   Burkard, R. E., and Fincke, V. *"On Random Quadratic Bottleneck Assignment Problems."* Math. Prog. 23 (1982): 227-232.

[2]   ———. *"Probabilistic Asymptotic Properties of Some Combinatorial Optimization Problem."* Manuscript. August 1982.

[3]   Erdös, P., and Rényi, A. *"On Random Graphs, I."* Publicationes Mathematicae Co., 1959, pp. 290-297.

[4]   Garey, M. R., and Johnson, D. S. *"Computers and Interactability: A Guide to the Theory of NP-Completeness."* Freeman, 1979.

[5]   Hochbaum, D. S. *"When are NP-hand Location Problems Easy."* Manuscript. November 1982, to appear in Annals of Operations Research, Vol. 1.

[6]   Karp, R. M. *"The Probabilistic Analysis of Some Combinatorial Search Algorithms."* In Algorithms and Complexity: New Directions and Recent Results. Academic Press, 1976.

[7]   Vercellis, C. *"A Probabilistic Analysis of the Set-Covering Problem."* Manuscript. University of Milan, December 1982. Annals of Operations Research, vol. 1 (to appear).

[8]   Weide, B. W. *"Statistical Methods in Algorithm Design and Analysis."* Thesis, CMU-CS-78-142. Carnegie-Mellon University, August 1978.

Annals of Discrete Mathematics 25 (1985) 211-238
© Elsevier Science Publishers B.V. (North-Holland)

NETWORK DESIGN WITH MULTIPLE DEMAND:
A NEW APPROACH

M. LUCERTINI

IASI–CNR and Dip. Informatica e Sistemistica Università Roma

G. PALETTA

CRAI Rende

The network design problem considered in this paper is an optimization problem, where a total demand cost function be minimized over budget constraints and network design variables, such as the topological structure and link capacities. The flow route selection, the separation of the flow in several commodities (with different source-sink couples), the indivisibility problems are not considered. The demand model proposed, take into account explicitly the uncertainty and the modifications of the demand vector during the operating period of the network. Therefore the demand cost function is a function of several feasible demand vectors instead of one. This network desing problem with multiple demand is formulated as a linear program. Some properties of the model are analyzed and a solution procedure is proposed both for the general case and for some simple demand models.

1. Introduction

Generally speaking a network synthesis or network design problem is one in which we are given a set of nodes and required to connect the nodes by arcs with sufficient capacities to ensure that specified network flow requirements can be met. In this paper we do not take into account the paths of the flows from sources to sinks, but only the

Mario Lucertini — Istituto di Analisi dei Sistemi ed Informatica del C.N.R. e Dipartimento di Informatica e Sistemistica dell'Università di Roma — Via Buonarroti, 12 - 00185 Roma

Giuseppe Paletta — Consorzio per la Ricerca e le Applicazioni di Informatica — Via Bernini, 5 — 87036 Rende (CS).

A first version of this paper appears as CRAI technical Report 82-14, 1982. Work performed by CRAI under contract n. 82.00024.93 "Progetto Finalizzato Trasporti".

possibility of sending a given amount of flow from one source to one sink. The problem is, in general, multi-commodity in the sense that each source-sink pair has a separate commodity to be transported through the network from the source to the sink and all the demands must be satisfied simultaneously.

In many applications the multicommodity model includes a fixed cost as well as a variable cost on the installation of capacity. From a formal point of view it is easy to incorporate such a cost in the model, but the model become a non linear problem (mixed integer linear programming problem). This extension is an order of magnitude more difficult to solve than the previous one and in particular, because of the non-convexity, the duality results cannot be utilized directly.

In the general formulation we have two conflicting goals: to minimize the global investment on the network (i.e. the total cost of the installation of arcs capacities) and to maximize the amount of satisfied demand. The most reasonable approach consist in deciding a fixed budget B and allocating the capacities such that a (weighted) sum of the demand flows will be maximum. This approach require a given demand vector. In fact the demand is, in almost all the practical applications, unknown and the best we can have, are bounds on the amount of demand in some specific nodes or in some subset of nodes and bounds on the total demand; we refer to this problem as a "multiple demand problem". A good approximation in the description of the demand set can be obtained by supposing that the demand vector must belong to a given polyhedron Ω in the demand space ($\Omega = \{d : Rd \leq r\}$). This generalization of the model leads to a more complex formulation where in particular we must define what we mean as "optimum". If G(N, A) is the network representing our problem where N is the set of nodes (sources, sinks and intermediate nodes) and A is the set of candidate arcs (arcs where we can increase the capacity, eventually all the possible arcs in the network), let $\Gamma(c)$ be the (convex) set of all the possible demand vectors that can be satisfied given a capacity vector $c(c_{ij}$ is the capacity of the arc (i, j) $\in A$, obviously the flow f_{ij} in the arc (i, j) must be $0 \leq f_{ij} \leq c_{ij}$).

The objective function proposed in this paper (to be maximized) is a suitable measure of the set ($\Omega \cap \Gamma(c)$), that is a measure of the set of possible demands that can be satisfied by the network or to minimize a suitable measure ϕ $(\Omega, \Gamma(c))$ of the set $(\Omega - \Gamma(c))$ (set of all demand vectors belonging to Ω but not to $\Gamma(c)$, i.e. set of unsatisfied demands). In this paper the network design problem with multiple demand is defined and formulated as a linear program (LP); some properties of the model are analyzed and a solution procedure based on the shortest path algorithm is pointed out. Furthermore the capacity expansion case (i.e. the case where we have an existing network with given arc capacities and a fixed budget and we must decide a capacity expansion such that we can meet "as well as possible" the demand requirements) is analyzed and a numerical example is given.

2. Model formulation and outline of the main results

2.1. The Model

Let us consider the problem with a unique source; it can be stated, as in the previous section, as

$$\min \phi \, (M, \Gamma(c))$$
$$c \geq 0$$

(P)

$$\sum_{(i, j) \in A} g_{ij} c_{ij} \leq B$$

with g_{ij} cost of an unitary increase of the capacity of arc (i, j).

The oriented Hausdorff distance [7] is utilized as a measure ϕ of the set $(\Omega - \Gamma(c))$.

Definition. *(Oriented Hausdorff distance). Given two sets* $H_1 \subseteq R^q$ *and* $H_2 \subseteq R^q$ *we define as "oriented Hausdorff distance"* ϕ (H_1, H_2) *between* H_1, *and* H_2 :

$$(H) \qquad \phi(H_1, H_2) = \max_{h^1 \in H_1} \; \min_{h^2 \in H_2} \; \|h^1 - h^2\|$$

where $\| \, . \, \|$ *is any norm in* R^n. ▲

A vector norm useful in many applications is the weighted sum (with positive weights) of the absolute value of the entries of the vector. In fact if we choose the weights equal to g_{ij} we obtain a model where we minimize the additional budget needed in order to satisfy the demand vector in the worst case.

In mathematical terms, let $G(N, A)$ be the graph representing the network with N set of nodes and A set of oriented arcs; let $D \subset N$ be the set of demand nodes, s the unique source, and c^O the initial capacities. The model M can be written as:

Model M

(M1)
$$\delta (c, d) = \min_{y, f} \sum_{(i, j) \in A} g_{ij} y_{ij}$$

$$0 \leq f \leq c^O + c + y$$

$$\sum_j f_{ij} - \sum_j f_{ji} = 0 \quad \forall i \in N - (D \cup \{s\})$$

$$\sum_j f_{ji} = d_i \qquad \forall i \in D$$

$$y \geq 0$$

(M2)
$$H(c) = \max_d \delta (c, d)$$

$$d \in \Omega = \{d : Rd \leq r\}$$

(M3)
$$z(B) = \min_c H(c)$$

$$\sum_{(i, j) \in A} c_{ij} g_{ij} \leq B$$

$$c \geq 0$$

The problem consists of finding c such that $z(B)$ is minimum. M1 corresponds to the minimization in (H), M2 to the maximization in (H), M3 to the minization problem P.

It is interesting to note that the whole model M can be written as a linear program (see the next section); in order to obtain this formulation it is necessary to consider the vertices of Ω, let $V = \{d^k : k \in K\}$ be the set of vertices of Ω, and K the set of vertex indices.

Problems M1 is a classical problem in network oprimization and can be solved by the Gomory and Hu algorithm [11, 14]. Many extension to multicommodity networks are also given (see [12, 15, 16, 17]). M2 can be viewed as a multicommodity problem with non-simultaneous flows.

2.2. *A Linear Programming Formulation of the Network Design Problem*

In order to formulate problem P as LP problem is useful to give the following lemma:

Lemma 1. *Given two bounded convex polyhedra H_1 and H_2 in R^q, let $V(H_1)$ be the set of the vertices of H_1, then*

$$\delta (H_1, H_2) = \delta (V(H_1), H_2). \quad \blacktriangleleft$$

The result is quite intuitive, the proof can be found in [8].

On the basis of this result M2 becomes a maximization problem on the whole set of the h vertex indices

$K = \{1, 2, ..., h\}$ of the set $\Omega = \{d : Rd \leq r\}$ (i.e. $V = \{d^1, d^2, ... d^h\}$).

We can therefore formulate M2 in the following way:

$$\max_{k \in K} \left\{ \min_{y^k \in S_k} \sum_{(i, j) \in A} g_{ij} y^k_{ij} \right\}$$

where S_k, $k \in K$ are the sets of values y^k defined by the constraints:

$$0 \leq f_{ij}^k \leq c_{ij}^o + c_{ij} + y_{ij}^k \qquad (i, j) \in A, k \in K$$

$$\sum_j f_{ij}^k - \sum_j f_{ji}^k = 0 \qquad i \in N-(D \cup \{s\}), k \in K$$

$$\sum_j f_{ji}^k = d_i^k \qquad i \in D, k \in K$$

$$c_{ij} \geq 0 \qquad (i, j) \in A$$

$$y_{ij}^k \geq 0 \qquad (i, j) \in A, k \in K$$

Where d_i^k is the i-th component of the k-th vertex of the polyhedron Ω.

Problem M3 is min-max-min problem and may be reformulated as an LP model.

Problem PH

min w
subject to:

(1) $\qquad \sum_{(i,j) \in A} c_{ij} g_{ij} \leq B$

(2) $\qquad c_{ij} + y_{ij}^k - f_{ij}^k \geq -c_{ij}^o \qquad \forall (i, j) \in A, \forall k \in K$

(3) $\qquad \sum_j f_{ij}^k - \sum_j f_{ji}^k = 0 \qquad \forall i \in N-(D \cup \{s\}), \forall k \in K$

(4) $\qquad \sum_j f_{ji}^k = d_i^k \qquad \forall i \in D, \forall k \in K$

(5) $\displaystyle\sum_{(i,j)\,\in A} g_{ij}y_{ij}^{k} \leq w$ $\qquad \forall k \in K$

(6) $c_{ij} \geq 0$ $\qquad \forall (i, j) \in A$

(7) $y_{ij}^{k} \geq 0, f_{ij}^{k} \geq 0$ $\qquad \forall (i, j) \in A, \forall k \in K$

2.3. Duality Results

The dual problem of (PH) is

$$\max\left\{ -Bt - \sum_{k \in K}\ \sum_{(i,\,j) \in A} c_{ij}^{o}u_{ij}^{k} + \sum_{k \in K}\ \sum_{i \in D} d_{i}^{k}v_{i}^{k} \right\}$$

subject to:

$-g_{ij}t + \displaystyle\sum_{k \in K} u_{ij}^{k} \leq 0$ $\qquad \forall (i, j) \in A$

$u_{ij}^{k} - g_{ij}z^{k} \leq 0$ $\qquad \forall (i, j) \in A, \forall k \in K$

$-u_{ij}^{k} + v_{j}^{k} - v_{i}^{k} \leq 0$ $\qquad \forall k \in K, \forall (i, j) \in A$

(PD) $\displaystyle\sum_{k \in K} z^{k} = 1$

$v_{s}^{k} = 0$ $\qquad \forall k \in K$

$u_{ij}^{k} \geq 0, \forall k \in K, \forall (i, j) \in A$

$t \geq 0$

$z^{k} \geq 0 \; \forall k \in K$

$v_{i}^{k} \, \forall i \in N, \forall k \in K$ unrestricted.

The meaning of the dual variables is the following:

t,   identified with the budget constraint (1)

v_i^k,   identified with the i–th node equation for the k-th demand vector (3, 4)

u_{ij}^k,   identified with the capacity constraint on arc (i, j) for the k-th demand vector (2)

z^k,   identified with the constraints (5).

The variables v_i^k (i $\in$ D) represent the cost for a unitary flow from s to i $\in$ D. The variables u_{ij}^k represent the value of a unitary increase (or decrease) in the initial capacity values.

Applying the optimal solutions orthogonality theorem, we obtain the following standard orthogonality conditions which are necessary and sufficient for optimality of primal and dual solutions:

$$t > 0 \rightarrow \sum_{(i,\, j) \in A} g_{ij} c_{ij} = B$$

$$u_{ij}^k > 0 \rightarrow c_{ij} + y_{ij}^k - f_{ij}^k = -c_{ij}^o \qquad \forall (i, j) \in A, \forall k \in K$$

$$z^k > 0 \rightarrow w - \sum_{(i,j) \in A} g_{ij} y_{ij}^k = 0 \qquad \forall\, k \in K$$

$$c_{ij} > 0 \rightarrow -g_{ij} t + \sum_{k \in K} u_{ij}^k = 0 \qquad \forall (i, j) \in A$$

$$y_{ij}^k > 0 \rightarrow u_{ij}^k - g_{ij} z^k = 0 \qquad \forall (i, j) \in A, \forall k \in K$$

$$f_{ij}^k > 0 \rightarrow -u_{ij}^k + v_j^k - v_i^k = 0 \qquad \forall (i, j) \in A, \forall k \in K$$

From the orthogonality conditions we may obtain substantial results which allow the simplification of the problem PD.

Remark that in the dual problem if $z^k = 0$ then $u_{ij}^k = 0 \ \forall (i, j) \in A$ and $v_i^k = 0 \forall i \in N$.

On the other hand $z^k > 0$ means that the alternative k is one of the "worst case" demand vectors (possibly the only one, the case of a unique "worst case" demand vector is analyzed separately in the sequel), in fact from the orthogonality conditions we obtain that if $z^k > 0$ then $w = \sum\limits_{(i, j) \in A} g_{ij} y_{ij}^k$, i.e. w is equal to the additional investment needed to satisfy the demand in the alternative k.

The following relations can be easily proved:

Lemma 2.

$$0 \leq \sum_{k \in K} u_{ij}^k \leq g_{ij} t^* \qquad \qquad \forall (i, j) \in A$$

$$t^* = \max_{(i, j) \in A} \left\{ \sum_{k \in K} u_{ij}^k / g_{ij} \right\} \leq 1$$

where t* indicate the optimal value of t. ◀

The constraints pointed out in the sequel refer to the optimal solution of the problem, but we omit the * to obtain a simpler notation.

Lemma 3. *There exists an optimal solution of problem* (PD) *such that* $u_{ij}^k = \max \left\{ v_j^k - v_i^k, 0 \right\} \ \forall k \in K.$ ◀

Proof. Given the values of (t, z^k, v_i^k), as $c_{ij}^0 \geq 0 \ \forall (i, j) \in A$, the optimal value of u_{ij}^k is the minimum value satisfying the constraints. But the only constraints bounding u_{ij}^k from below are $u_{ij}^k \geq v_i^k - v_i^k$ and $u_{ij}^k \geq 0$. ◀

The optimal solution of (PD) is in general not unique, in fact it is easy to prove that if $c_{ij}^0 = 0, \ \forall (i, j) \in A$, there exist optimal solutions of problem PD such that $\sum\limits_{k \in K} u_{ij}^k$ can assume all the values in the interval:

$$\left[\left(\sum_{k \in K} v_j^k - \sum_{k \in K} v_i^k \right), g_{ij} t \right]$$

From the orthogonality constraints we obtain that if $\sum\limits_{k \in K} u_{ij}^k \neq g_{ij} t$ then $c_{ij} = 0$; but assuming $u_{ij}^k = v_j^k - v_i^k$ then if $(v_j^+ - v_i^+) < (g_{ij} t)$ then $c_{ij} = 0$, where $v_j^+ = \sum\limits_{k \in K} v_j^k$. The dual of problem M1 can be written as:

$$\max \left\{ - \sum_{(i, j) \in A} c_{ij} g_{ij} - \sum_{(i, j) \in A} c_{ij}^o u_{ij} + \sum_{i \in D} d_i v_i \right\}$$

(D M1) $v_j - v_i - u_{ij} \leq 0$ $\qquad\qquad \forall (i, j) \in A$

$\qquad\quad v_s = 0$

$\qquad\quad 0 \leq u_{ij} \leq g_{ij}$ $\qquad\qquad \forall (i, j) \in A$

In the section 3 it will be shown that this formulation corresponds to the formulation of the general model with a unique demand vector.

2.4. *Main Results*

In order to introduce the results and the algorithmic approach proposed in this paper, let us first formulate problem M in a slightly different way.

Problem M *for an increase* ΔB *of budget*

Given $c \geq 0$, find $\gamma \geq -c$ such that $\delta(c+\gamma, d^k) \leq b$, $\forall k$ and b will be minimum, subject to the constraint $g\gamma = \Delta B$.

$$\lambda\ (\Delta B) = \min_{\gamma} b$$

(MC)

$$\delta\ (c +\gamma,\ d^k) \leq b \qquad\qquad \forall k \in K$$

$$g\gamma = \Delta B$$

$$\gamma \geq -c \qquad\qquad \blacktriangleleft$$

The total budget utilized is obviously equal to $g(c +\gamma)$.

In practice it is not necessary for small values of ΔB to consider all the vertices in V, but only a subset of vertices called "critical vertices".

Definition (Critical vertex). *Given c, a vertex d^k of Ω is said to be critical if*

$$\delta(c, d^k) \geq \delta(c, d^j) \qquad\qquad \forall j \in K$$

Let $K(c)$ indicate the set of critical vertices for the given c. $\qquad \blacktriangleleft$

If for some given optimal c there exist a unique critical vertex, i.e. $|K(c)| = 1$, (by $|K(c)|$ we denote the cardinality of the set $K(c)$) the optimal investment strategy consists in investing on the arcs with $y_{ij} > 0$ in problem M1 for a maximum amount equal to $(g_{ij}\ y_{ij})$ for each arc and until a new vertex becomes critical. In this case for every ΔB units of budget invested, the objective function is reduced by exactly ΔB units and a relatively simple solution algorithm exists.

This case will be analyzed in section 4.

If for a given c there exist two or more critical vertices, the optimal investment strategy consists of investing the given amount of budget and possibly in reallocating the budget previously invested on the various arcs, in order to minimize the distance from all the critical vertices.

This case is more complex and leads to a large scale linear programming problem for each value of B. An approximate approach can be introduced as follows:

APPROXIMATE APPROACH

i)   Divide the total budget in parts equal to ΔB and formulate the problems as M_Δ.

ii)   At each step invest ΔB on the most convenient arc.

iii) At each step, after the increase of $(\Delta B/g_{ij})$ capacity units on the selected arc (i, j), verify for each couple of arcs if a reallocation of ΔB units of budget is feasible and decrease the objective function or the number of critical vertices, if this is the case, perform the substitution and iterate until no more reallocation is convenient.

The interesting point is that, with a suitable choice of "the most convenient arc" and with a suitable definition of the reallocation procedure, an algorithm based on the previous approximate approach has found in all the test problems and the applications a solution different from the optimal one by at most ΔB for all the budget values. Furthermore the algorithm converges in a finite number of steps (at each step or reallocation the distance with respect to at least one critical vertex decreases by a finite quantity) and it is considerably more efficient than an algorithm solving M parametrically with respect to B.

Let $z_A(B, \Delta B)$ be the solution obtained through the approximate algorithm for a budget B with an increase of budget at each step equal to ΔB.

Theorem 1.

$$z_A(B, \Delta B) - z(B) \leq |A| \cdot \Delta B$$

Proof. Let c be the optimal capacity vector in problem **M** (corresponding to $z(B)$) and $\tilde{c}$ the capacity vector obtained through the approximate algorithm. No reallocation procedure of ΔB units of budget is possible to reduce the distance from at least one of the critical vertices. Each reallocation can affect the objective function for at most ΔB units. But the set of feasible capacity vectors is a convex polyhedron the result follows. ◀

Remark that the reallocation procedure is the core of the approximation algorithm. In fact when applied it guarantee the approximation. If we require the approximation only for a given budget B, it is not needed to apply the reallocation procedure in each step of the algorithm, but only for the given B.

The dual of (M_Δ) can be written as:

$$\max \left\{ - \left(\sum_{(i,\,j) \in A} (c_{ij}\; g_{ij} + \Delta B)\, t - \right. \right.$$
$$\left. \left. \sum_{(i,\,j) \in A} c_{ij}^{o} \left(\sum_{k \in K} u_{ij}^{k} \right) + \sum_{k \in K} \sum_{i \in D} d_{i}^{k}\, v_{i}^{k} \right\} \right.$$

subject to the same constraints as (PD).

An interesting extension of the model can be obtained by introducing the cost of excess capacity for each alternative. Let equation (1) of PH be an equality constraint and equations (5) be modified as follows:

(5) $$\sum_{(i,\,j) \in A} g_{ij} (2y_{ij}^{k} + c_{ij} - f_{ij}^{k}) =$$

$$= B + \sum_{(i,\,j) \in A} g_{ij} (2y_{ij}^{k} - f_{ij}^{k}) \leq w \qquad \forall\, k \in K$$

each equation is obtained by the sum of the expansion costs plus the excess capacity costs evaluated for a simpler presentation at the same unitary cost g_{ij}.

If now we consider B instead of a given value, a variable of the problem, the optimal value B* such that w is minimized can be considered in some sense the "optimal" investment on the network. In fact a smaller B increase the (worst case) expansion costs and a greater B increase the (worst case) excess capacity costs.

2.5. *Set of Critical Veritices*

A crucial point in the previous model concern the size of the set of critical vertices. In fact only a subset of V (set of vertices of Ω) can be considerate to become critical.

Definition. *A vertex* $d^i \in V$ *is said to be a "dominated vertex" if there exists a convex combination* $\bar{d}$ *of a subset of* $V - \{d^i\}$ *such that* $d^i \leq \bar{d}$; *otherwise* d^i *is said to be a "nondominated vertex". (In particular* d^i *is a dominated vertex if there exist* $d^j \in (V - \{d^i\}$ *such that* $d^i \leq d^j$). ◄

It is easy to verity that the "nondominated vertices" are the vertices V' obtained as follows:

$$S = (\underset{d \in \Omega}{\cup} \{d' : 0 \leq d' \leq d\})$$

$V(S)$ set of vertices of S

$$V' = V(S) \cap V$$

The problem is that, in the worst case, $|V'|$ can grow exponentially with the number of constraints defining the set Ω.

Nevertheless if ΔB is small enough to guarantee that all the optimal basis for all the critical vertices remain the same and the set of critical vertices remain the same, then the behaviour of all the minimal distances from the critical vertices also is the same.

In practice the exponential growth of $|V'|$ is only a theoretical problem and it is unlikely that we must deal with a large member of vertices; in practical experiments the number of critical vertices does not exceed the dimension of the demand space $|D|$.

3. Unique critical demand vector

If $|K(c)| = 1$ we have results of particular interest from an algorithmic point of view.

The above condition means that there is a "dominant vertex" for the given capacity values c, i.e. the worst case increase of investment corresponds to a single vertex. In particular this is generally the case for $c^o + c = 0$ and in the approximate procedure where we invest at each step a fixed amount ΔB on a single arc.

In problem (M_Δ) only the constraints associated to critical vertices are active, therefore it can also be written as:

$$\lambda (\Delta B) = \min_\gamma b$$

$$\delta (c + \gamma, d^k) \le b \qquad\qquad \forall\, k \in K(c)$$

$$g\gamma = \Delta B$$

$$\lambda \ge -c$$

$$\gamma \text{ such that } K(c+\gamma) = K(c)$$

Let us indicate by Π_k the set of capacities such that the vector d^k is the unique critical vector of the problem.

Lemma 4. *If $(c^o + c) \in \Pi_k$ for some $k \in K$ and $w > 0$ then the optimal value of the dual variable t is $t^* = 1$.* ◀

Proof. If $(c^o + c) \in \Pi_k$ then $|K(c)| = 1$, hence all the z^k but one are equal to zero, let $\hat{k}$ be the index such that $z^{\hat{k}} = 1$; if $w > 0$ then it does exist at least one arc (i, j) such that $y_{ij}^{\hat{k}} > 0$, therefore from the orthogonality constraints we obtain $g_{ij} = u_{ij}^{\hat{k}}$, but also $u_{ij}^{\hat{k}} \le g_{ij} t \le g_{ij}$, and the result follows. ◀

Utilizing the previous result (and eliminating the index $\hat{k}$ and the constant term in the objective function) the dual problem can be written as follows:

$$\max_{u,\,v} \left\{ - \sum_{(i,\,j) \in A} c_{ij}^o u_{ij} + \sum_{i \in D} d_i v_i \right\}$$

$$v_j - v_i \le u_{ij} \qquad\qquad \forall\, (i, j) \in A$$

(MU)

$$v_s = 0$$

$$0 \le u_{ij} \le g_{ij} \qquad\qquad \forall\, (i, j) \in A$$

(MU) gives the optimal solution for all capacities $(c^o+c+\gamma)$ such that $(c^o+c+\gamma) \in \Pi_k$.

For any given value u, problem MU is equivalent to the problem of finding the shortest path tree on the network with arc weights equal to u_{ij} from s to the demand nodes. In particular if $c^o = 0$, the arc weights are equal to g_{ij}. The optimal value of the objective function is equal to $(\sum_{i \in D} d_i p_i\,(u))$, where $p_i(u)$ is the value of the shortest path from s to $i \in D$ given the values of all the u_{ij}.

Given u, MU can be easily solved through any shortest path algorithm, on the other hand given v, MU can also be easily solved ($u_{ij} = $ max $(0, v_j - v_i)\ \forall\, (i, j) \in A$). By eliminating the constraint $u \ge 0$ MU can be written

$$\max \left\{ \sum_{i \in N-D} \left(\sum_{j:(i,\,j) \in A} c^o_{ij} - \sum_{j:(j,\,i) \in A} c^o_{ji} \right) v_i + \right.$$

(MUA)

$$\left. + \sum_{i \in D} \left(d_i + \sum_{j:(i,\,j) \in A} c^o_{ij} - \sum_{j:(j,\,i) \in A} c^o_{ji} \right) v_i \right\}$$

subject to

$$v_j - v_i \le g_{ij} \qquad\qquad \forall\, (i, j) \in A$$

$$v_s = 0$$

The dual of MUA is

$$\min_{(i,\,j)\in A} g_{ij}f_{ij}$$

(DMUA)

$$Zf = a$$

$$f \geq 0$$

Where Z is the incidence matrix of the graph, f_{ij} represents the flow on the arc (i, j). DMUA can be solved as a transportation problem, from both nodes with a negative value of the a_i and sources ($a_s =$

$$= - \sum_{i \in D} d_i$$), to nodes with a positive value of a_i.

The a_i 's represent the available or required quantities. The cost for transferring a flow unit from source i, to destination j, is equal to the shortest path value from i to j on the graph with the arcs weights equal to g.

Solution of MUA gives in general a lower bound of the MU solution; if all values $(v_j - v_i)$ corresponding to non zero values of c_{ij}^o are non-negative , then the solution is also optimal (for small values of c_{ij}^o, this is the case) otherwise the value of the DMUA-solution plus

$$\sum_{(i,\,j)\in A} c_{ij}\, \eta(v_j - v_i), \text{ where}$$

$$\eta(v_j - v_i) = \begin{cases} v_j - v_i & \text{if } v_j \leq v_i \\ \\ 0 & \text{otherwise} \end{cases}$$

is an upper bound on the optimal solution.

If the approximation is satisfactory, we can take this solution as MU solution, otherwise we can utilize standard technique to find the optimal solution.

4. A sufficient condition for reducing problem M to a problem with a unique demand vector

Given a set of arc capacities $(c^o + c)$ and the corresponding set of critical vertices $K(c)$, let $\bar{d}$ be the demand vector defined as follows:

$$\bar{d}_i = \min_{k \in K(c)} d_i^k$$

and let $\bar{c}$ be the minimum cost increase of arc capacities needed to satisfy $\bar{d}$; $\bar{c}$ can be obtained as the solution of problem M1 with $d = \bar{d}$.

It is easy to prove the following result:

Theorem 2. *If* $K(c + \bar{c}) = K(c)$ *and* c *is optimal for* **M** *with a budget* $B = gc$, *then* $(c + \bar{c})$ *is also optimal for problem* **M** *(with a budget* $\bar{B} = g(c + \bar{c}))$, *i.e. corresponds to the maximal decrease in the objective function.*

Proof. The investment $g\bar{c}$ produces a decrease of the same amount $g\bar{c}$ in all the $\delta(c, d^k)$, $k \in K(c)$, hence if the set of critical nodes does not change the solution is optimal.   ◀

Let d^{min} be the following demand vector:

$$d_i^{min} = \min_{k \in K} d_i^k$$

let c^{min} be the solution of problem M1 with $d = d^{min}$ and f the corresponding flow values; the following corollary is a direct consequence of theorem 2.

Corollary 1. c^{min} *is the optimal solution of problem* **M** *with a budget equal to* (gc^{min}).

Corollary 1 leads to a preprocessing procedure utilizing the algorithm with a unique demand vector.

In practice we can generate a new problem with demand vectors given by $(d^k - d^{min})$, initial arc capacities given by $(c^0 + c^{min} - f)$ and a budget equal to $(B - gc^{min})$.

5. Approximate algorithm

Stating the following definitions:

— B^{max} total budget available;

— ΔB is the units of budget to be invested at each iteration;

— G is the matrix of the expansion cost of the capacity;

— δ is the future investment needed to meet the demand vector;

The algorithm proposed is the following:

INPUT: $G(N, A)$, c^0, B^{max}, G, ΔB, D

OUTPUT: c

begin

    $B: = 0$, $c: = c^0$

until $B = B^{max}$ *repeat*

begin

        $\delta_m: = 0$, $B: = B + \Delta B$, $K(c): = \{0\}$

    for $k \in K$ *do*

    begin

procedure DELTA (c, d^k, δ, u)

if ($\delta > \delta_m$)   *then* K(c): $= \{k\}$, δ_m: $= \delta$

$\qquad\qquad$ *else if* ($\delta = \delta_m$) *then* K(c): $=$ K(c)$+ \{k\}$

end;

if ($\delta_m = 0$)   *then* STOP

$\qquad\qquad$ *else* ε_m : $= 0$, IA: $= 0$

for $(i; j) \in$ A *do*

begin

procedure EPSILON $((i, j), \varepsilon, u_{ij}, g_{ij}, K(c))$

if ($\varepsilon > \varepsilon_m$)   *then* IA: $= \{(i,j)\}$, ε_m: $= \varepsilon$

$\qquad\qquad$ *else if* ($\varepsilon = \varepsilon_m$) IA: $=$ IA$+ \{(i, j)\}$

end

if (|IA|: 1)   *then* c_{ij}: $= c_{ij} + \Delta B/g_{ij}$ $(i, j) \in$ IA

$\qquad\qquad$ *else*

$\qquad\qquad\qquad$ *begin*

$\qquad\qquad\qquad\qquad$ g_m: $=$ B^{max}

$\qquad\qquad\qquad\qquad$ *for* $(i, j) \in$ IA *do*

$\qquad\qquad\qquad\qquad\qquad$ *begin*

$\qquad\qquad\qquad\qquad\qquad\qquad$ *if* ($g_{ij} < g_m$)

$\qquad\qquad\qquad\qquad\qquad\qquad\qquad$ *then* g_m: $= g_{ij}$, (s, r): $= (i, j)$

$\qquad\qquad\qquad\qquad\qquad$ *end*

$\qquad\qquad\qquad\qquad$ c_{sr}: $= c_{sr} + \Delta B/g_{sr}$

$\qquad\qquad\qquad$ *end*

procedure REALLOCATION ((i, j), (r, s), ΔB, g, c, D, δ_m, K(c))

$if (\delta_m) = 0)$ STOP

end

end

DELTA (c, d^k, δ, u)

INPUT (c, D), OUTPUT (δ, u)

    — Solve M1

    — Calculate the corresponding value of $\delta := \delta(c, d^k)$

    — Calculate the dual variables u_{ij}^k $k \in K(c)$, (i, j) $\in$ A

EPSILON ((i, j), ε, u_{ij}, g_{ij}, K(c))

INPUT (u_{ij}, g_{ij}, K(c)), OUTPUT (ε_{ij})

    — Calculate $\varepsilon_{ij} = \sum_{k \in K(c)} u_{ij}^k / g_{ij}$

REALLOCATION ((i, j), (r, s), ΔB, g, c, D, δ_m, K(c))

begin

NREA: = 0

    until NREA = 0 *repeat*

    begin

    for (i, j) $\in$ A *do*

    for (r, s) $\in$ $\left\{A-(i, j)\right\}$ *do*

$$c_{ij}^1 := c_{ij} - \Delta B/g_{ij};$$

$$c_{rs}^1 := c_{rs} + \Delta B/g_{rs};$$

$$\bar{\delta} := 0; \bar{K}(c) = 0;$$

    for $k \in K$

      begin

        Procedure DELTA (c, d^k, δ, u)

        if $(\delta > \bar{\delta})$ *then*
                begin

                    $\bar{\delta} := \delta$

                    $\bar{K}(c) := \{k\}$

              end

        else
          if $(\delta = \bar{\delta})$ *then* $\bar{K}(c) := \{\bar{K}(c) + k\}$

      end

    if $((\bar{\delta} < \delta_m)$ or $((\bar{\delta} = \delta_m)$ and $(|K(c)| > |\bar{K}(c)|))$

        then $c_{ij} := c_{ij}^1; c_{rs} := c_{rs}^1; \delta_m := \bar{\delta}; \text{NREA} := 1$

        else $\text{NREA} := 0$

   end

end

Example

Let be given the following network (initial capacities are set to zero) and demand vectors (d^1, d^2).

$$d^1 = \begin{pmatrix} d_2^1 \\ d_1^1 \end{pmatrix} = \begin{pmatrix} 2 \\ 1 \end{pmatrix}, \ d^2 = \begin{pmatrix} d_2^2 \\ d_1^2 \end{pmatrix} = \begin{pmatrix} 1 \\ 2 \end{pmatrix}$$

The problem is to invest a budget value equal to 22 minimizing the future investment needed to meet the "worst" (in terms of cost) demand vector.

Let

$$\Delta B = 2$$

According to theorem 1, the first step consist in a suitable increase of the capacities of the arcs in order to meet the demand vector

$$d^{min} = \begin{pmatrix} 1 \\ 1 \end{pmatrix}$$

Such objective is achived by setting:

$$c_{si}^{min} = 1, \ c_{c2}^{min} = 1, \ c_{12}^{min} = 0$$

which requires an investment value equal to 12. We can now generate a new problem with demand vectors given by:

$$d^1 = \begin{pmatrix} d_2^1 \\ d_1^1 \end{pmatrix} = \begin{pmatrix} 1 \\ 0 \end{pmatrix}, \ d^2 = \begin{pmatrix} d_2^2 \\ d_1^2 \end{pmatrix} = \begin{pmatrix} 0 \\ 1 \end{pmatrix}$$

Initial capacities of the arcs are:

$$c_{S1} = 0, \quad c_{S2} = 0, \quad c_{12} = 0$$

and the available budget is equal to 10.

Consequently, the incremental costs needed to meet each demand vector are:

$$\delta(c, d^1) = 8$$

$$\delta(c, d^2) = 4$$

Iteration 1. Since d^1 is the critical demand, we invest on the $(s, 2)$.

We obtain

$$c_{S1} = 0, \quad c_{12} = 0 \quad c_{S2} = 1/4$$

$$\delta(c, d^1) = 6$$

$$\delta(c, d^2) = 4$$

$$B = 8$$

The invested budget has been allocated in an optimal way, therefore the reallocation procedure does not lead to any improvement.

Iteration 2. The critical demand is still d^1, so that we invest on the $(s, 2)$ arc:

$$c_{S1} = 0, \quad c_{12} = 0, \quad c_{S2} = 1/2$$

$$\delta(c, d^1) = 4$$

$$\delta(c, d^2) = 4$$

B = 6

Iteration 3. The critical demand are d^1 and d^2, it's convenient to invest on the (s, 1) arc because this will produce a decrease for both $\delta(c, d^1)$ and $\delta(c, d^2)$

$$c_{S1} = 1/2, \ c_{12} = 0, \ c_{S2} = 1/2$$

$$\delta(c, d^1) = 3$$

$$\delta(c, d^2) = 2$$

$$B = 4$$

Iteration 4. Now the critical demand is d^1, thus it is convenient to invest on (1, 2)

$$c_{S1} = 1/2, \ c_{12} = 1/3, \ c_{S2} = 1/2$$

$$\delta(c, d^1) = 1$$

$$\delta(c, d^2) = 2$$

$$B = 2$$

Iteration 5. The critical demand is d^2, so we invest on (s, 1).

$$c_{S1} = 1, \ c_{12} = 1/3, \ c_{S2} = 1/2$$

$$\delta(c, d^1) = 1$$

$$\delta(c, d^2) = 0$$

$$B = 0$$

The reallocation procedure, not used in the previous iterations since it did not lead to any impovment, procedures now:

$$c_{S1} = 1, \quad c_{12} = 2/3, \quad c_{S2} = 1/4$$

$$\delta(c, d^1) = 1/2$$

$$\delta(c, d^2) = 0$$

By using again the reallocation procedure, we obtain finaly:

$$c_{S1} = 1, \quad c_{12} = 1, \quad c_{S2} = 0$$

$$\delta(c, d^1) = 0$$

$$\delta(c, d^2) = 0$$

References

[1] G. Gallo: *Lower planes for the network design problem.* ISI – Pisa, Ott. 1981.

[2] G. Gallo: G. Sodini: *Concave cost minimization on networks.* EJOR, 1979.

[3] G. Gallo: *A new branch and bound algorithm for the network design problem.* IEI – Pisa, 1981.

[4] D.S. Johnson, J. K. Lenstra, A. H. G. Rinnooy Kan: *The complexity of the network design problem.* Networks, 1978.

[5] M. Luss: *Operations research and capacity expansion problems: a survey.* Operations Research, 1982.

[6] D. E. Bell: *Regret in decision making under uncertainty.* Operations Research, 1982.

[7] Hogan: *Point to set maps in mathematical programming.* UCLA, febbraio 1971.

[8] M. Lucertini, G. Paletta: *Espansione e riallocazione di capacità su una rete logistica in condizioni di incertezza.* AIRO Journal (to appear).

[9] M. Lucertini: *Bounded rationality in long term planning: a linear programming approach.* Metroeconomica, vol. 24, n. 3 1982.

[10] M. Lucertini, G. Paletta: *Politiche di investimento su una rete di flusso in presenza di domande multiple: analisi di alcuni casi.* Giornate AIRO, Napoli, 1983.

[11] T. C. HU: *Combinatorial algorithms.* Addison Wesley, 1982.

[12] R. E. Gomory, T.C. HU: *Multiterminal network flows.* SIAM App. Math., 1961.

[13] R. E. Gomory, T.C. HU: *Synthesis of a Communication Network.* SIAM, 1964.

[14] R. E. Gomory, T. C. HU: *An application of generalized linear programming to network flows.* SIAM App. Math. 1962.

[15] J. L. Kennington: *A survey of linear cost multicommodity network flows.* Op. Res. 1978.

[16] J. A. Tomlin: *Minimum cost multicommodity network flows.* Op. res. 1966.

[17] M. Minoux: *Optimum synthesis of a network with nonsimultaneous multicommodity flow requirements.* Annals of Discrete Mathematics, vol. 11, 1981.

Annals of Discrete Mathematics 25 (1985) 239-254
© Elsevier Science Publishers B.V. (North-Holland)

HOW TO FIND LONG PATHS EFFICIENTLY

B. MONIEN

Universität Paderborn

We study the complexity of finding long paths in directed or undirected graphs. Given a graph $G = (V, E)$ and a number k our algorithm decides within time $0(k! \cdot |V| \cdot |E|)$ for all $u, v \in V$ whether there exists some path of length k from u to v. The complexity of this algorithm has to be compared with $0(|V|^{k-1} \cdot |E|)$ which is the worst case behaviour of the algorithms described up to now in the literature. We get similar results for the problems of finding a longest path, a cycle of length k or a longest cycle, respectively.

Our approach is based on the idea of representing certain families of sets by subfamilies of small cardinality. We also discuss the border lines of this idea.

1. Introduction

In this paper we study the problem of determining a path of length k in a directed or undirected graph $G = (V, E)$. This problem is closely related to the longest path problem and to some other problems which we will describe later. By a 'path' we always mean a 'simple path' (see [5]), i.e. we do not allow that a vertex appears on a path more than once. Without this restriction (i.e. by allowing a vertex to appear more than once on a path) or for graphs without cycles the problems is wellknown (see [10]) to be solvable in polynomial time whereas the problem of determining a simple path of length k, k arbitrary, is NP-complete. One can consider this problem as a single-source and single-destination problem (i.e. as the problem to decide for fixed $u, v \in V$ whether there exists a path of length k from u to v) or as the more general problem to decide for all $u, v \in V$ whether there exists a path of length k from u to v. We will study here the second approach. That is we want to compute a matrix $D^{(k)} = (d_{ij}^{(k)})$, $1 \leq i, j \leq n$, where $d_{ij}^{(k)}$ is equal to some path from i to j of length k, if such a path exists, and $d_{ij}^{(k)}$ is equal to some special symbol λ, if there exists no path from i to j of length k.

The straightforward algorithm which enumerates all sequences of length k+1 solves the problem within time $0(|V|^{k+1})$. It has to consider for every pair of nodes (i, j) and for any sequence $u_1,..., u_{k-1}$ of nodes

$$i, u_1 , \ldots u_{k-1} , j$$

whether $i \rightarrow u_1 \rightarrow u_2 \rightarrow \ldots \rightarrow u_{k-1} \rightarrow j$ holds. We get the slightly better time bound $0(|V|^{k-1} \cdot |E|)$ if we take into account that we have only to consider nodes u_1 with $(i, u_1) \in E$. In the case k=2 the problem can be solved also by squaring the adjacency matrix of G (which leads to an estimation $0(|V|^{\alpha})$ with $\alpha < 3$, see [1]). To our knowledge no algorithm solving this problem for arbitrary k in less than $0(|V|^{k-1} \cdot |E|)$ time has been published.

The algorithm of Latin Multiplication which is described above all in the literature (see [10])for solving this problem computes for $1 \leq p \leq k$ (or for p = 1, 2, 4, ..., k, respectively) the matrices containing all paths of length p. This algorithm also has a worst case complexity of the above order.

We have seen that for any fixed k we have a polynomial time algorithm to solve this problem but its computational behaviour is terrible if the numbers k and $|V|$ are not very small. In this paper we will describe an algorithm whose behaviour is much better and which will solve the problem for small k rather efficiently.

Theorem 1: Let G = (V, E) be any graph and let $k \in N$. The matrix $D^{(k)}$ (G) can be computed within time $0(C_k \cdot |V| \cdot |E|)$, where $C_k = k!$.

Note that we have replaced the time bound $0(|V|^{k-1} \cdot |E|)$ by $0(C_k \cdot |V| \cdot |E|)$, $C_k = k!$. Estimations of this kind can also be found for other problems. We want to mention here the vertex cover problem ([12], time bound $0(2^{\mu/2} + |E|)$, where μ is the cardinality of the solution) and the feedback vertex set problem for undirected graphs (unpublished result of the author, time bound $0(2^{\mu} \cdot (\log\mu)^{\mu} \cdot |V| \cdot |E|)$ where again μ is the cardinality of the solution). Our new algorithm allows to compute the solution for instances, e.g. $|V| = 2o$ and k = 7, which were outside the computational practicability before.

The estimation of our algorithm depends on the one side on k and on the other side on $|V|$ and $|E|$. The dependence on k (i.e. $C_k = k!$)

is not optimal. The reader will notice that we do not estimate very sharp in our proof. We will discuss this topic again at the end of section 2. On the other hand the behaviour in $|V|$ and $|E|$ seems to be close to optimal, i.e. an improvement of this behaviour would lead to improved algorithms also for other wellknown problems. Note that already in order to compute the matrix of all paths of length 2 we need $0(|V| \cdot |E|)$ time (at least this is our present knowledge) if we are not willing to use one of the algorithms for fast matrix multiplication. A similiar observation can be made when we are faced with the problem of deciding whether there exists a cycle of length k in the given graph G. This problem can be solved by computing first the matrix $D^{(k-1)}(G)$ and then comparing $D^{(k-1)}(G)$ with E, i.e. for every $(i, j) \in E$ we look whether there exists a path of length k-1 form j to i. Therefore we can compute in time $0(C_{k-1} \cdot |V| \cdot |E|)$ whether a graph $G = (V, E)$ has a cycle of length k. The problem of determining whether a graph has a triangle (i.e. the case k=3) has been studied carefully, since for undirected graphs there exists a n^2-reduction from the problem of determining a shortest cycle to the problem of determining a triangle, [8]. Also for the problem of determining a triangle only algorithms of time complexity $0(|V| \cdot |E|)$ and the algorithms for fast matrix multiplication are known, [8]. Therefore it is likely that the time bound $0(|V| \cdot |E|)$, which holds for any fixed k, is rather sharp.

Note that as a simple corollary of Theorem 1 we have proved above the following theorem.

Theorem 2: Let $G = (V, E)$ be any graph and let $k \in \mathbb{N}$. We can decide within time $0(C_{k-1} \cdot |V| \cdot |E|)$, $C_k = k!$, whether G has a cycle of length k and compute such a cycle if it exists.

We will show in Section 4 that we can use Theorem 1 also to find a longest path and a longest cycle efficiently (for finding a longest cycle we can apply our method only if the graph is undirected).

Theorem 3: Let $G = (V, E)$ be any graph. We can compute a longest path of G within time $0(c_{\mu+1} \cdot |V| \cdot |E|)$, $C_\mu = \mu!$, where μ is the length of the longest path of G.

Theorem 4: Let G = (V, E) be an undirected graph. We can compute a longest cycle of G within time $0(C_{2\mu-1} \cdot |V| \cdot |E|)$, $C_\mu = \mu!$, where μ is the length of the longest cycle of G.

It was shown before, [7], that a longest cycle in an arbitrary graph G = (V, E) can be found within time $0(|V|^\mu \cdot |E|)$ where μ is the length of the longest cycle of G. Both problems, determining a longest path as well as determining a longest cycle, are well known and well studied and are important in many applications (see [13]).

Before we start to prove our theorems we want to give an idea about the method we use. We feel that this method is quite general and should have further applications.

Let a graph G = (V, E), V = $\{1,..., n\}$, and a number k be given. The first consideration is very simple. We start from the set of edges and then we compute successively for all p, $2 \leq p \leq k$, and for all i, j ϵ V all the paths of length p from i to j. If we have already computed for all i, j ϵ V all the paths from i to j of length p, then we can use this information to compute the paths of length p +1 by Latin Multiplication (see [10]). This approach is very time consuming since the number of paths of length p can be very large. We overcome this difficulty by considering instead of the family of all paths of some given length between two nodes some subfamily which we really need in order to compute finally paths of length k. We call this subfamily a representative for the family of all paths.

We will describe this idea in the next section and we will also give a close upper bound for the cardinality of the optimal representatives (Theorem 5). In section 3 we show how representatives can be computed efficiently and prove theorem 1. There is still a rather large gap between the cardinality of the representatives we get in section 3 and the optimal ones. In section 4 we prove theorem 3 and theorem 4.

2. The use of representatives

Our first step is to consider instead of paths (i.e. sequence of nodes) the sets of nodes lying on a path, i.e. we don't distinguish between paths running over the same set of nodes. From now on we will use in our proof only these sets. In an implementation of our algorithm one should encode such a set as a sequence of nodes which form a path in G in order to have really paths available when the algorithm stops.

Let us set for $1 \le i,j \le n, 0 \le p \le n-1$

$$F_{ij}^{p-1} = \left\{ U \in P_{p-1}(n) \mid U \text{ occurs as the set of inner nodes on a path of length } p \text{ from } i \text{ to } j \right\}.$$

Here $P_{p-1}(n)$ denotes the family of all subsets of $\{1,..., n\}$ of cardinality p-1. Let us consider as an example the graph G given by figure 2.1.

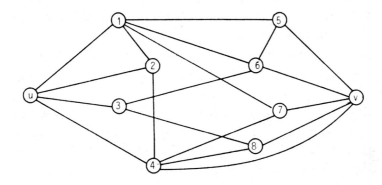

Figure 2.1: The graph G

Then $F_{uv}^2 = \left\{ \{2,4\}, \{1,5\}, \{1,6\}, \{1,7\}, \{3,6\}, \{3,8\}, \{4,8\}, \{4,7\} \right\}$.

Now let us define the notion of a representative. Let $q \in N$ with $0 \le q < n$ and let F be any family of sets over $\{1,..., n\}$. A q-representative $\hat{F}$ for F is defined in such a way that if we consider any set $T \subset \{1,..., n\}$ of cardinality at most q and ask whether F contains a set U with $T \cap U = \emptyset$ then we get the correct answer also by looking only through $\hat{F}$.

Definition: Let F be a family of sets over $\{1,..., n\}$ and let $q \in N$, $0 \le q < n$. A subfamily $\hat{F} \subset F$ is called a q-representative of F if the following condition holds:

For every $T \in P_{\le q}(n)$, if there exists some $U \in F$ with $T \cap U = \emptyset$ then there exists also some $\hat{U} \in \hat{F}$ with $T \cap \hat{U} = \emptyset$.

Let us consider again the above example. $\hat{F}: = \{\{2,4\},\{1,5\}\}$ is a 1-representative for F^2_{uv}. Since $\hat{F}$ contains two disjoint sets for any $T \subset \{1,..., n\}$ with $|T| = 1$ the family $\hat{F}$ contains a set $\hat{U}$ with $T \cap \hat{U} = \emptyset$. Because of the analogous reason $\hat{F}: = \{\{2,4\},\{1.5\},\{3,6\}\}$ is a 2-representative for F^2_{uv} and it is not difficult to see that $\hat{F}: = \{\{2,4\}, \{1,5\},\{3,6\},\{1,7\},\{3,8\},\{4,8\}\}$ is a 3-representative for F^2_{uv}.

We have said that we will use the idea of the representative to compute the matrix $D^{(k)}$. Let $u,v \; \epsilon \; V$ be two nodes. We have to decide whether there exists a path of length k from u to v. What do we have to know about the sets F^{k-2}_{ij}, $i,j \; \epsilon \; V$?

A path from u to v of lenght k consists of an edge $\{u,i\} \; \epsilon \; E$, $i \neq v$, and a path from i to v of length k-1 from i to v which does not contain u.

$$u \qquad i \quad . \; . \; . \qquad v$$
$$\overline{\hspace{3cm}}$$
$$\text{length k-1}$$

Therefore it is sufficient to know for every $i \; \epsilon \; V$ whether there exists a path from i to v of length k-1 not containing u. This information is given by a 1-representative for F^{k-2}_{iv}. We can formulate this simple observation in the following way:

Assume that we know 1-representatives for F^{k-2}_{ij}, $1 \le i, j \le n$. Then we can compute 0-representatives for F^{k-1}_{ij}, $1 \le i,j \le n$.

Note that for any family F a 0-representative $\hat{F}$ of F is empty iff F is empty and it has to contain only one arbitrary set from F if F is not empty.

We can easily generalize the above observation and get the following lemma which we will call the main lemma because of its importance for this paper.

Main lemma: Let p,q be numbers with $0 \le p < n$ and $1 \le q \le n$. Assume that we know q-representatives for F^p_{ij}, $1 \le i,j \le n$, but not necessarily the sets F^p_{ij} itself. Then we can compute (q-1)-representatives for all the families F^{p+1}_{ij}, $1 \le i,j \le n$.

We can use the idea of the main lemma by computing first (k-2)-representatives for F^1_{ij}, $1 \le i,j \le n$, and then (k-3)-representatives for F^2_{ij}, $1 \le i,j \le n$,....., until we reach 0-representatives for F^{k-1}_{ij}, $1 \le$

i,j $\leq$ n. We will show in the next section that we can do this computation efficiently. Closely related with the complexity of this computation is the maximum number of sets which may belong to a representative. Therefore we define

$$\alpha\ (p,\ q,\ n) = \max_{F \subset P_p(n)} \quad \min \left\{ |\hat{F}| \ ; \hat{F} \text{ is a q-representative for } F \right\}$$

It is remarkable that we know this function explicitely. Results from [4, 9] imply that $\alpha(p,\ q,\ n) = \binom{p+q}{p}$ for $n \geq p+q$. Note that no proper subset of $F = P_p\ (p+q)$ is a q-representative of F and therefore $\alpha(p,\ q,\ n) \geq \binom{p+q}{p}$. In order to prove the other direction we have to introduce some new definitions.

Let $F \subset P_p\ (n)$. A set $T \subset \left\{ 1,...,\ n \right\}$ is called a hitting set of F if $U \cap T \neq \emptyset$ for all $U \in F$.

F is called q-minimal if for every $U \in F$ the family $F - \left\{ U \right\}$ has a hitting set of cardinality q which is not a hitting set of F.

It is clear that $\hat{F} \subset F$ is a q-representative of F iff every hitting set of $\hat{F}$ of cardinality at most q is also a hitting set of F.
This implies that every family $F \subset P_p(n)$ has a q-representative which is q-minimal.

It was conjectured in [3] and shown in [4] and [9] (see also [2]), that every family $F \subset P_p\ (n)$ which is q-minimal contains at most $\binom{p+q}{p}$ sets. Therefore we get the following theorem:

Theorem 5: $\alpha(p,\ q,\ n) = \binom{p+q}{p}$ for $n \geq q + q$.

This theorem does not imply that we can compute a q-representative with cardinality $\leq \binom{p+q}{p}$ efficiently. The method which we will use in the next section leads only to q-representatives of cardinality $\sum_{i=1}^{q} p^i$. Therefore we do not think that the constant $C_k = k!$ in our Theorem 1 is close to be optimal. Note that a lower bound for C_k using the method of representatives is given by

$$\sum_{p=1}^{k-1} \alpha(p,k-p-1,n) = \sum_{p=1}^{k-1} \binom{k-1}{k-p-1} = \sum_{r=0}^{k-2} \binom{k-1}{r} = 2^{k-1} - 1.$$

It was already noticed in [3] that $\alpha(p, q, n) \leq \sum_{i=1}^{q} p^i$

The author realized the connections between the work of [3, 4, 9] and the work presented here only during the last stage of preparing this paper.

3. Proof of theorem 1:

We want to compute the matrix $D^{(k)} = (d_{ij}^{(k)})$, where $d_{ij}^{(k)}$ is some special path from i to j of length k, if it exists, and $d_{ij}^{(k)} = \lambda$ otherwise. As we described in the introduction we have to compute (k-p-1)-representatives for all the sets F_{ij}^p, i,j ϵ V, $1 \leq p \leq k - 1$.

Actually we define trees whose nodes are labelled with the sets from F_{ij}^p such that the family of all the sets which occur as node labels in this tree form a (k-p-1)-representative of F_{ij}^p. The tree structure enables us to do the computations, described by the main lemma, efficiently. We will call such a tree a (k-p-1)-tree for F_{ij}^p.

Definition: Let $F \subset P_p$ (n) be a family of sets. Let q be some natural number. A q-tree for F is a p-nary node labelled and edge labelled tree of height at most q which satisfies the following conditions:

(i)   Its nodes are labelled with sets from F or with the special symbol λ. Its edges are labelled with elements from $\{1,..., n\}$.

(ii)   If a node is labelled with some set U ϵ F and if its depth is less than q, then it has p sons and each of the p elements of U occurs as a label of one of the edges connecting this node with its sons.

(iii) If a node is labelled with the special symbol λ or if its depth is equal to q, then it has no sons.

(iv)   Between the labels of the nodes and the edges the following relation holds: For any node ξ of this tree, if $E(\xi)$ is the set of elements from $\{1,..., n\}$ occurring as edge labels on the path from the root of this tree to ξ, then either label $(\xi) \epsilon$ F and label $(\xi) \cap E(\xi) = \emptyset$ or label $(\xi) = \lambda$ and there exists no U ϵ F with U $\cap$ $E(\xi) = \emptyset$.

As an example (see figure 3.1) we want to describe a 3-tree for the set F_{uv}^2 which we considered in the introduction, i.e. for $F = F_{uv}^2 =$

$\{\{2,4\},\{1,5\},\{1,6\},\{1,7\},\{3,6\},\{3,8\},\{4,8\},\{4,7\}\}$. Note that for $0 \leq q \leq 2$, the first q levels of this tree form a q-tree for the family F.

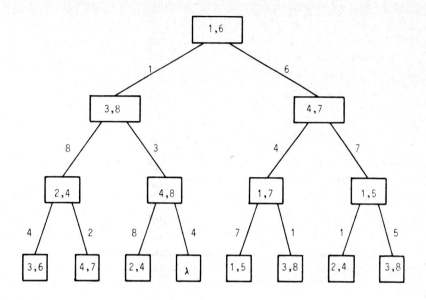

Figure 3.1: A 3-tree for the set F_{uv}^2

Lemma 1: Let $F \subset P_p$ (n) be a family of sets, let q be some natural number and let B be some q-tree for F. Then the family $\hat{F}$ consisting of all sets which occur as node labels in B form a q-representative of F. Furthermore we can decide for every $T \epsilon P_q$ (n) in $0(p \cdot q)$ steps whether there exists some $U \epsilon F$ with $T \cap U = \emptyset$ and compute such a U if it exists.

Proof: We will prove the second assumption first. Consider the following algorithm:

procedure Disjoint-Set (ξ: node of B; T: element of $P_{\leq q}$ (n))
begin
   if label (ξ) $\neq$ λ *and* label (ξ) $\cap$ T $\neq$ $\emptyset$ *then*
   begin
      Let $\hat{\xi}$ be some son of ξ such that the edge
      from ξ to $\hat{\xi}$ is labelled with some element

a ϵ label $(\xi) \cap T$;
call Disjoint-Set $(\hat{\xi}, T)$
  end else
  If label $(\xi) = \lambda$ *then* write (There exists no $U \epsilon F$ with $U \cap T = \emptyset$)
  else if label $(\xi) \cap T = \emptyset$ *then* write $(U = $ label (ξ) fullfills $U \in F$
    and $U \cap T = \emptyset$);
end;

Initially we call this procedure with Disjoint-set (root of B, T) and we have to show that it always produces the correct output. We observe three facts:

1.) If the algorithm finds a node ξ with Label $(\xi) \cap T = \emptyset$ then clearly $U = $ label (ξ) has the property that $U \epsilon F$ and $U \cap T = \emptyset$ since every node label either is the special symbol λ or a set belonging to F.

2.) Now assume that the algorithm reaches a node ξ with label (ξ) $= \lambda$. Let $E(\xi)$ be defined as in the definition of the q-tree. This definition implies that there exists no $U \epsilon F$ with $U \cap E(\xi) = \emptyset$. But because of our algorithm $E(\xi) \subset T$ and therefore there exists no $U \epsilon F$ with $U \cap T = \emptyset$.

3.) There still is to show that always one of the write-statements is reached. If ξ is a node of depth $\hat{q}$, $\hat{q} < q$, then either we reach a write-statement or we call the procedure again with some node $\hat{\xi}$ of depth $\hat{q} + 1$. If ξ is a node of depth q, then $|E(\xi)| = q$ and since on the other hand $E(\xi) \subset T$ and $|T| = q$ we can conclude that in this case $E(\xi) = T$. Therefore if label $(\xi) \neq \lambda$ then label $(\xi) \cap T = \emptyset$ and we reach a write-statement since the condition of the while-statement is not fullfilled.

We have shown now that our algorithm computes a set $U \epsilon F$ with $T \cap U = \emptyset$ if such a set exists. The computation needs $0(q \cdot p)$ steps, since the number of calls of the procedure is bounded by the depth of the tree B (and this depth is bounded by q) and since during every call two sets of size p have to be compared (which needs $0(p)$ steps).

Thus our second assumption is proved. The first assumption follows directly from the above consideration since the above algorithm computes for every $T \in P_{\leq q}(n)$ some set $U \in F$ with $T \cap U = \emptyset$ if such a set U exists. Furthermore this set U occurs as a node label of tree B and therefore it belongs to $\hat{F}$. Thus $\hat{F}$ is a q-representative of G. $\qquad\square$

Being a p-nary tree of depth at most q, B has at most $(p^{q+1}-1)/(p-1)$ nodes and therefore the cardinality of the representative $\hat{F}$ is bounded by $(p^{q+1}-1)/(p-1)$.

Now we want to show that if q-trees for all the sets F_{ij}^p $1 \leq i,j \leq n$, are given, then we can compute efficiently (q-1)-trees for the sets F_{ij}^{p+1}.

Lemma 2: Let $0 \leq p \leq n$, $1 \leq q \leq n$. Assume that q-trees $B_{i,j}^p$ for F_{ij}^p, $1 \leq i,j \leq n$, have already been computed. For $u,v \in \{1,..., n\}$ we can compute a (q-1)-tree for F_{uv}^{p+1} in time $O(q \cdot (p+1)^q \cdot \text{degree}(u))$, where degree (u) is the degree of u in the graph G.

Proof: We compute the node labels and the edge labels of the (q-1)-tree B for F_{uv}^{p+1} level-wise, i.e. we compute first the label of the root of B and the labels of the edges leaving the root. After having computed the labels for all the nodes of depth i and all the edges connecting nodes of depth i with nodes of depth i+1, we determine the labels for the nodes of depth i+1.

Now let ξ be some node of depth i+1. Let $E(\xi) \subset \{1,..., n\}$ be the set of edge labels on the path from the root to ξ. Note that all these edge labels have already been computed. We have to find a set $U \in F_{uv}^{p+1}$ with $U \cap E(\xi) = \emptyset$ (if it exists).

Note that every path from u to v of length p+2 consists of one edge $(u,w) \in E$, $w \neq v$, and a path from w to v of length p+1 which does not contain the node u. Therefore there exists a set $U \in F_{u,v}^{p+1}$ with $E(\xi) \cap U = \emptyset$ if and only if there exists some $w \in \{1,... n\} - \{v\}$ with $(u,w) \in E$ and some $\hat{U} \in F_{wv}^p$ with $\hat{U} \cap (E(\xi) \cup \{u\}) = \emptyset$. But for every $w \in V$ with $(u, w) \in E$ we can decide because of lemma 1 in $O(p \cdot q)$ steps whether there exists a set $\hat{U} \in F_{wv}^p$ with $\hat{U} \cap (E(\xi) \cup \{u\}) = \emptyset$. Since we do this computation at most degree (u) times, we can compute one node label in time $O(p \cdot q \cdot \text{degree}(u))$. Computing the labels of the edges leaving this node takes no additional time. The lemma follows since B has at most $\frac{(p+1)^q}{q}$ nodes. $\qquad\square$

Note that F_{ij}^O contains exactly the empty set if $(i,j) \in E$ and it is the empty family if $(i,j) \notin E$. Therefore a q-tree for F_{ij}^O has the form $\emptyset$, if $(i, j) \in E$ and the form λ, if $(i, j) \notin E$.

We have to compute the matrix $D^{(k)}$ which we get because of Lemma 1, if we know all the 0-trees for F_{ij}^{K-1}, $1 \le i,j \le n$. We start from the $(k-1)$-trees for F_{ij}^O, $1 \le i,j \le n$, (which we don't have to compute since these 'trees' are given by the set of edges E) and then we compute successively the $(k-2)$-trees for $F_{i,j}^1$, $1 \le i,j \le n$, the $(k-3)$-trees for F_{ij}^2, $1 \le i,j \le n$, and so on. All these computations can be performed because of Lemma 2 within the time.

$$c \cdot \sum_{p=1}^{k-1} p^{k-p} \cdot (k-p) \cdot |V| \cdot |E| \le c \cdot (k-1) \cdot \sum_{p=1}^{k-1} p^{k-p} \cdot |V| \cdot |E|.$$

It can be shown easily by induction that $\sum_{p=1}^{k-1} p^{k-p} \le (k-1)!$ for $k \ge 5$. Thus we have proved Theorem 1.

Theorem 1: Let $G = (V, E)$ be any graph and let $k \in N$. The matrix $D^{(k)}(G)$ can be computed within time $0(C_k \cdot |V| \cdot |E|)$, where $C_k = k!$.

4. Proof of theorem 3 and theorem 4

It is clear that we find a longest path by computing successively the matrices $D^{(1)}, D^{(2)}, D^{(3)},...$ until we reach for the first time a matrix $D^{(\ell)}$ whose entries are all equal to λ. Then the matrix $D^{(\ell-1)}$ has some entry which is not equal to λ and this entry is a longest path. The computation of $D^{(1)}, D^{(2)},..., D^{(\ell)}$ needs no more time than the computation of only $D^{(\ell)}$. This is true since when we have computed some $D^{(k)}$ and have to compute $D^{(k+1)}$ then all the trees which have been constructed while computing $D^{(k)}$ can be used and have to be enlarged by one level.

Theorem 3: Let $G = (V, E)$ be any graph and let μ be the length of the longest path of G. We can compute a longest path of G within time $0(C_{\mu+1} \cdot |V| \cdot |E|), C_\mu = \mu!$.

The application of our method for computing a longest cycle is not so obvious. We are able to do so only for undirected graphs. In the case of undirected graphs there is some relationship between the length of the longest path and the length of the longest cycle. It was shown in [11] that in any k-connected graph with a longest path of length ℓ the length of the longest cycle is at least $\dfrac{2k-4}{3k-4} \cdot \ell$. We will not use this result here but use some simple lemma.

Lemma 3: Let G be an undirected graph and let Δ be the diameter of G. Suppose there exists a cycle C with $|C| \geq 2 \cdot \Delta + 2$. Then there exists also a cycle $\hat{C}$ with $\dfrac{1}{2} |C| < |\hat{C}| < |C|$.

As usually the diameter denotes the maximum distance in G.

Before we prove the lemma we want to show that it gives a sharp estimation. Consider the graph G given by figure 4.1.

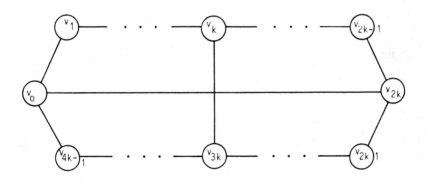

Figure 4.1: The graph G

This graph has a cycle of length 4k, its diameter is $\Delta = k + 1$ and besides its Hamiltonian cycle it has only cycles of length $2k + 1$ and $2k + 2$.

Proof of lemma 3:

Let C be a cycle with $|C| \geq 2\Delta + 2$. Set $k = \left\lfloor \dfrac{|C|}{2} \right\rfloor$. Then $k > \Delta$ holds. Let a,b be two nodes on C such that both paths from a to b on C have length at least k. Let P_1, P_2 be the two paths on C from a to b. Then $|P_1|$, $|P_2| \geq k$. Let P be a shortest path from a to b in G, $|P| \leq \Delta < k$.

We have to consider two cases.

(i)   Except for the endpoints P and P_1 (or P and P_2, respectively) are vertex-disjoint. Then $\hat{C} = PP_1$ (or $\hat{C} = PP_2$, respectively) fulfills the conditions of the lemma.

(ii) Besides a,b the path P contains some further node from P_1 and some further node from P_2. Then we can assume that there exist d and e such that the path P has the from described by figure 4.2 and the following conditions hold:

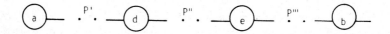

Figure 4.2: Partition of the path P

d belongs to $P_1 - \{a\}$, e belongs to $P_2 - \{b\}$, P′ contains no inner point from P_2 and P″ contains no point from P_1 or from P_2.

Let $P_{11}, P_{12}, P_{21}, P_{22}$ be the subpaths of P_1, P_2 defined by d and e (see figure 4.3).

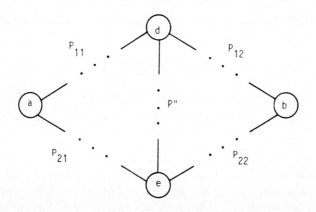

Figure 4.3: Cycle C, path P″

Since P is the shortest path from a to b we know that $|P''| \leq |P_{12}|$ and $|P''| \leq |P_{21}|$ holds. Let C_1 denote the cycle $P_{11}P''P_{21}$ and let C_2 denote the cycle $P_{12}P''P_{22}$. Then $|C_1| + |C_2| = |C| + |P''| > |C|$ and $|C_1| < |C|$ (because of $|P''| < |P_{12}|$) and $|C_2| < |C|$ (because of $|P'''| < |P_{21}|$). Therefore the longer one of the two cycles C_1, C_2 fulfills the conditions of the lemma.

$\square$

We use this lemma in order to compute a longest cycle. We can assume that the graph is biconnected (otherwise we compute the biconnected components, this needs time $0(|E|)$, see [1,5]). Then we compute the diameter Δ of G and for two nodes a,b with distance Δ we determine two vertex-disjoint paths from a to b, i.e. we compute a cycle of length k, $k \geq 2\Delta$ (this computation needs time $0(|V| \cdot |E|)$, see [1,5]. Then we compute successively D^ℓ for $\ell = k, k + 1,...$. By comparing D^ℓ with E we check whether there exists a cycle of length $\ell + 1$. We stop when we have reached for the first time some $D^{2\ell-1}$ such that there exists no cycle of length μ for $\ell < \mu \leq 2\ell$. Because of Lemma 3 we know that in this case the length of the longest cycle is equal to ℓ.

Theorem 4: Let G be an undirected graph and let μ be the length of the longest cycle of G. We can compute a longest cycle of G within time $0(C_{2\mu-1} \cdot |V| \cdot |E|)$, $C_\mu = \mu!$.

Acknowledgement: The author wants to thank R. Schulz, E. Speckenmeyer and 0. Vornberger for the many discussions we had during the preparation of this paper.

References

[1]   Aho, A.V., J.E. Hopcroft and J.D. Ullman: The Design and Analysis of Computer Algorithms, Addison-Wesley, 1974
[2]   Berge, C: Graphs and Hypergraphs, North Holland-American Elserier, 1973
[3]   Erdös, P, and T. Gallai: On the Minimal Number of Vertices Representing the Edges of a Graph, Publ. Math. Inst. Hung. Ac. Sc. (Mag. Tud. Akad.) 6(1961), 181 - 203
[4]   Erdös, P., A. Hajnal and J. Moon: A problem in Graph Theory, Math. Notes, Am. Math. Monthly 71(1964), 1107 - 1110
[5]   Even, S.: Graph Algorithms, Pitman Publishing Limited, 1979
[6]   Garey, M.R. and D.S. Johnson: Computers and Intractability, Freeman and Company, 1979

[7]   Hsu, W., Y. Ikura and G.L. Nemhauser: A polynomial algorithm for maximum weighed vertex packings on graphs without long odd cycles, Math. Progr. 2o(1981, 225 - 232

[8]   Itai, A. and M. Rodeh: Finding a Minimum Circuit in a Graph, Proc. 1977 ACM Symp. Theory of Computing, 1 - 10

[9]   Jaeger, F. and C. Payan: Détermination du nombre maximum d'âretes d'un hypergraphe T-critique, C.R. Acad. Sc. Paris 273(1971), 221 - 223

[10] Kaufmann, A. : Graphs, Dynamic Programming and Finite Games, Academic Press, 1967

[11] Locke, S.C.: Relative Lengths of Paths and Cycles in k-Connected Graphs, J. Comb. Th. B 32(1982), 206 - 222

[12] Monien, B. and 0. Vornberger: unpublished paper

[13] Roy, B. : Algebre moderne et théorie des graphes, tome 1,2, Dunod, 1969

Annals of Discrete Mathematics 25 (1985) 255-280
© Elsevier Science Publishers B.V. (North-Holland)

COMPACT CHANNEL ROUTING OF MULTITERMINAL NETS

M. SARRAFZADEH and F. P. PREPARATA

Coordinated Science Laboratory and Department of Electrical Engineering, University of Illinois, Urbana, IL 61801.

In this paper we describe a novel technique for solving the channel routing problem of multiterminal nets. The layout is produced column-by-column in a left-to-right scan: the number t of used tracks satisfies the bound $\delta \leq t \leq \delta + \alpha (0 \leq \alpha \leq \delta\text{-}1)$, where δ is the density of the problem. The technique behaves equivalently to known optimal methods for two-terminal net problems. For a channel routing problem with C columns and n nets, the algorithms run in time $O(C\log n)$ and produces layouts that are provably wirable in three layers.

Keywords: layout techniques, channel routing problem, knock-knee layout mode, multiterminal nets, two-terminal nets, multilayer wiring.

1. Introduction

A general two-shore channel routing problem (GCRP) consists of two parallel rows of points, called *terminals,* and a set of nets, each of which specifies a subset of terminals to be (electrically) connected by means of *wires.* The goal is to route the wires in such a way that the channel width is as small as possible.

The following concepts are illustrated in Figure 2, below. As is customary, we view a channel of width t as being on a unit grid with grid points (x, y), where both x and y are integers, with $0 \leq y \leq t+1$ and arbitrary x. The horizontal lines are called *tracks* and the vertical lines *columns.* A vertex (x, y) of this grid at either $y = 0$ or $y = t+1$ is a

This work was supported in part by the Semiconductor Research Corporation under Contract 83-01-035 and by the National Science Foundation under Grant MCS-81-05552.

terminal; in particular, $(s_j, 0)$ is a *lower* (or *entry*) terminal and $(t_i, t + 1)$ is an *upper* (or *exit*) terminal. A *wire* is a subgraph of this grid whose edges are segments connecting adjacent vertices in the grid. A *net* N is an ordered pair of (not simultaneously empty) integer sequences $((s_1,..., s_k), (t_1,..., t_h))$; thus, N contains lower terminals $s_1,..., s_k$ and upper terminals $t_1,..., t_h$. If $k + h > 2$, then we speak of a *multi-terminal net,* as distinct from a *two-terminal net;* the reason for this distinction is that channel routing of two-terminal nets is much simpler and better understood than the corresponding multiterminal net problem.

A solution to a GCRP must have, for each net, a graph on the grid that contains a path between any two terminals of that net. Note also that no two nets may share the same terminal. We shall adopt the layout mode known as "knock-knee" [RBM, L, BB], where *no two wires share and edge of the grid,* but two wires may cross at a vertex or may both bend at that vertex (see Figure 1). In this mode, two distinct nets can share only a finite number of points, thereby reducing crosstalk between nets.

In the channel there is a fixed number (two or more) of *conducting layers,* each of which is a graph isomorphic to the channel grid. These layers are (ordered and) placed one on top of another, and contacts between two distinct layers (*vias*) can be made only at grid points. If two layers are connected at a grid point, no layer inbetween can be used at that grid point.

We shall use the terms "layout" and "wiring" with the following distinct technical connotations (as in [PL]).

Definition 1. A *wire layout* (or simply *layout*) for a given GCRP is a subgraph of the layout grid, each of whose connected components corresponds to a distinct net of the GCRP, in the knock-knee mode.

Notice that we can, without loss of generality, restrict ourselves to connected subgraphs which are trees, called *wire-trees* (see Figure 2 (b)). (Each non-tree graph can be replaced by one of its tree subgraphs on the same set of terminals.)

Definition 2. Given a wire layout consisting of wire-trees $w_1,...w_n$, a *wiring* is a mapping of each edge of wire-tree w_i (for $i = 1, 2,..., n$) to a conducting layer with vias established at layer changes.

An optimal layout of a given GCRP is a layout that uses the least possible number d of tracks. A simple-minded (and optimistic) lower bound to d can be readily established as follows.

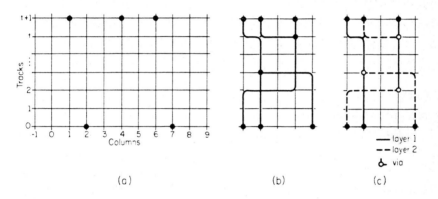

Figure 1. Illustration of the basic construct in the knock-knee mode.

Figure 2.   (a)   Specification of the terminal of a net [net ((2,7), (1, 4, 6))];
         (b)   wire layout of two nets (two wire-trees are shown);
         (c)   wiring of the layout of Figure (b).

Consider a GCRP $\eta = \{N_1, ..., N_n\}$, where $N_i = ((s_1^i, ..., s_{k_i}^i),$ $(t_1^i, ..., t_{h_i}^i))$, and let $\ell_i = \min(s_1^i, t_1^i)$ and $r_i = \max(s_{k_i}^i, t_{h_i}^i)$. The interval $[\ell_i, r_i]$ represents an obvious lower bound to the horizontal track demand raised by N_i, since a terminal in column ℓ_i must be connected to a terminal in column r_i. In other words N_i is replaced by a fictitious two-terminal net N_i^* (whose two terminals may belong to the same track). We now consider the channel routing problem $\eta^* = \{N_1^*, ..., N_n^*\}$, and use standard methods to obtain its density δ (i.e., the maximum number of two-terminal nets which must cross any vertical section of the channel). It is clear that δ is a lower bound for the minimum number of horizontal tracks, and we call δ the *essential density* of the GCRP.

Two methods [BP] [B] have been recently proposed for the GCRP, which use at most 2δ tracks. The methods are inherently different, but both produce the layout track-by-track; the method in [B] used exactly 2δ tracks, while the method in [BP] frequently results in more economical realizations.

In this paper we shall illustrate a method that produces the layout column-by-column (analogously to the "greedy router" of [RF]), and uses $\delta + a$ tracks, where $0 \leq \alpha \leq \delta\text{-}1$.

The paper is organized as follows. In Section 2 we describe and prove the correctness of the wire layout algorithm, beginning with a systematic version that uses $2\delta\text{-}1$ tracks, and then introducing natural modifications leading to potentially simpler layouts. We also show that the method, when applied to a collection of two-terminal nets, produces a result equivalent to the one of the optimal method of Preparata-Lipski [PL]. Finally, in Section 3 we show that the obtained layouts are wireable in three layers.

2. Wire layout algorithm

Before describing our proposed wire layout algorithm, we examine a simpler version thereof, which uses exactly $2\delta\text{-}1$ tracks. This simpler technique will provide the intuitive background for the method; the final algorithm will be a refinement of this version.

2.1 An Algorithm Achieving $t = 2\delta\text{-}1$.

For any integer c, the interval $(c, c+1)$ is called a *vertical section* (the vertical strip comprised between two columns). We say that a net $N = ((s_1,..., s_k), (t_1, ..., t_h))$ is *upper-active in $(c, c+1)$* if $t_1 \leq c \leq t_h$, and *lower-active in $(c, c+1)$* if $s_1 \leq c \leq s_k$; N is *active* in $(c, c+1)$ if it is both upper-active and lower-active in $(c, c+1)$. Any vertical line $x = x_0$, $x_0 \epsilon (c, c+1)$ for $\min(s_1, t_1) \leq c \leq \max(s_k, t_h)\text{-}1$ cuts N in at least one point; each intersection of N with $x = x_0$ identifies a *strand* of N at x_0.

The invariant (and the central feature) maintained by the algorithm is the following:

Property 1. In the vertical section $(c\text{-}1, c)$ each upper-active net has a strand lying *above* a strand of any lower-active net; similarly, a lower-active net has a strand lying *below* a strand of any upper-active net.

If this property holds, then the layout of the column is straightforward. Indeed, if $c = s_p^j = t_q^i$ (i.e., column c has an entry terminal of net N_j and an exit terminal of net N_i), by Property 1 there is a strand σ_i of N_i lying above a strand σ_j of N_j; thus we connect s_p^j to σ_j and t_q^i to N_i by means of nonoverlapping vertical wires (see Figure 3).

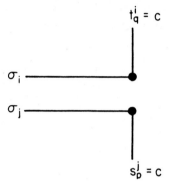

Figure 3. Column layout

We shall now assume that the layout of the given GCRP η uses tracks -t, -t + 1,..., t. Denoting by J the set of indices of the nets of η, we define a function $\sigma : J \to \{1,2,..., t\}$. We then recognize that Property 1 holds if we algorithmically guarantee the following (specialized) invariant:

Property 2. In the vertical section (c-1, c) each net N_m upper-active in (c-1, c) has a strand at $y = \sigma(m)$; each net N_m lower-active in (c-1, c) has a strand at $y = -\sigma(m)$. This means that if N_m is active in (c-1, c), it has a symmetric pair of strands at $y = \sigma(m)$ and $y = -\sigma(m)^{(1)}$ (Figure 4).

In other words, if Property 2 is maintained the strands of a net N_m can occur only in two fixed tracks ($y=\sigma(m)$ or $y = -\sigma(m)$) symmetric about the track $y = 0^{(1)}$. The algorithm we are about to describe is claimed to maintain Property 2.

(1) Thus, by convention, track $y = 0$ is not used. Later, we shall see that the strands (of the same net) at $y = +1$ and $y = -1$ can be made to coincide.

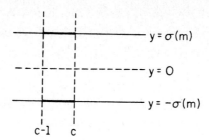

Figure 4. The two symmetric strands of N_m active at c.

For convenience, in the GCRP statement, a net will be represented by a tree as in Figure 5. Suppose we display all the members of $\eta = \{N_1,..., N_n\}$ each as in Figure 5, in the correct vertical alignment. For a given column c, we cut a vertical slice $[c-\varepsilon, c+\varepsilon]$, $0 < \varepsilon < 1$, and retain only the net fragments containing a terminal (at most two): this yields the *column state* (state(c)), i.e., the layout requirement of column c.

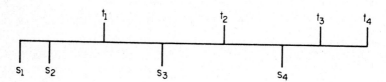

Figure 5. Representation of a multiterminal net in GCRP statement.

The 20 different possible states are shown in Figure 6 (where $|$ denotes "empty", and "↑" denotes "trivial", i.e., a two-terminal net with $s_1 = t_1 = c$). For a column c, let $d_L(c)$ and $d_R(c)$ be the local densities of the problem in the vertical sections (c-1, c) and (c, c+1), respectively. With reference to a left-to-right scan, we say that c is a density increasing column (d.i.c.) if $d_L(c) < d_R(c)$, and is a density decreasing column (d.d.c.) if $d_L(c) > d_R(c)$. With this definition, the column states are readily classified as in Figure 6 as d.i. (density increasing), d.d. (density decreasing) and d.p. (density preserving). Hereafter, *a terminal will be labeled with the index of the net to which it belongs.*

We begin by considering the d.p. columns. States $\top$, $\perp$, $+$ and $\frac{\perp}{\top}$ are readily handled, as shown in Figure 3 (or in a trivial variant thereof). So we must consider states $\dashv_\Gamma$ and $\neg L$. Here N_i terminates and N_j begins. The two nets N_i and N_j can be concatenated to form a *run of* nets. The transition between two nets of the same run can be handled very simply. In either of the cases illustrated in Figure 7, we assign to N_j the same track(s) assigned to N_i to the left of c, and Property 2 is maintained.

We now give a less informal description of the handling of a d.p. column. Specifically, here and hereafter, a *right bend* is a layout construct of the types "⌐" or "Γ", whereas a *left bend* is one of the constructs "⌐" or "L".

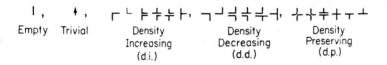

		Density	Density	Density
Empty	Trivial	Increasing	Decreasing	Preserving
		(d.i.)	(d.d.)	(d.p.)

Figure 6. Possible column states.

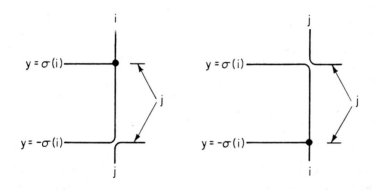

Figure 7. Handling of a density-preserving column.

Available is an array describing the function σ, and a priority queue Q of the available tracks in the set $\{y = i \mid i = 1,..., \delta\}$. In the following subroutine, LAYOUT D.P.C. (u, ℓ; c), u and ℓ are respectively the names of the nets having an upper (exit) or lower (entry) terminal in column c. Of course, either u or ℓ (or both) may be equal to **Λ**, (the index of the empty net); in which case any "connect" operation involving the empty net is void.

procedure LAYOUT–D.P.C. (u, ℓ; c)
begin if state(c) ϵ { $\top$, $\perp$, $+$, $\frac{\perp}{\top}$ } *then* connect upper terminal to
    y = σ(u) and lower terminal to y = -σ(ℓ)
  else begin
       r : = net starting at c;
       if state (c) = $\neg$ $\llcorner$ *then*
          begin connect lower terminal to y = σ(ℓ) and y = -σ(ℓ);
              σ(r) : = σ (ℓ);
              connect upper terminal to σ(r) with left bend
        end
       else
          begin connect upper terminal to y = σ(u) and y = -σ(u);
              σ(r) : = σ(u);
              connect lower terminal to -σ(r) with right bend
        end
      end
end

By forming runs of nets and handling them as shown in Figure 7, we are effectively partitioning the channel into *blocks* of contiguous columns, so that changes of density occur only at columns separating adjacent blocks. The preceding discussion gives us the following result.

Lemma 1: Inside eack block the set of tracks used remains fixed.

We now consider the handling of d.d. columns, with the assumption (to be substantiated later) that a terminating net has at most two disconnected strands in tracks above and below y = 0. Recall that at a d.d. column one or two nets terminate. Suppose, at first, that just one net terminates (states $\neg$, $\lrcorner$, $\dashv$, $\frac{\perp}{\top}$, $\frac{\perp}{\top}$). States $\neg$, $\lrcorner$, and $\dashv$ are handled in a straightforward manner, by connecting both strands of the net to the appropriate terminal. There remain states $\frac{\perp}{\top}$ and $\frac{\perp}{\top}$, of which we just need to consider $\frac{\perp}{\top}$ (the other case being handled symmetrically). Referring to Figure 8, let N_i be the terminating net and N_j be the continuing net. If $\sigma(j) > \sigma(i)$, then the termination of N_i is straightforward (Figure 8a). If $\sigma(j) < \sigma(i)$, then N_i cannot be terminated at column c. Thus we extend both strands of N_i to the closest d.i. or empty column e to the right of c. If e is an empty column, then the two strands of N_i are connected at e in a straightforward manner (Figure 8b).

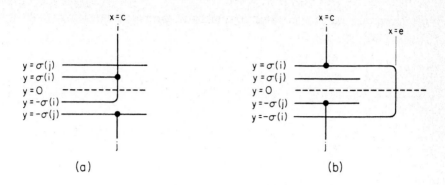

Figure 8. Handling of a density-decreasing column c; N_i is the terminating net.

When e is a d.i. column, the operation of splicing of the two strands of N_i at e will be considered later in connection with the layout of a d.i. column. When two nets terminate at c (state(c) = ⊣), then one net is spliced at c (as N_i in Figure 8a) and the other is extended (as N_j in Figure 8b). We shall make use of an integer parameter EX, which denotes the number of nets being extended to the right of the current column c.

The preceding discussion is formalized in the subroutine LAYOUT-D.D.C. (u, ℓ; c; EX); this subroutine makes use of the priority queue Q of the available tracks (only positive ordinates), with the usual notations "Q ⇐" (add to Q) and "⇐ Q" (extract from Q), and, of a priority queue P of the extended nets.

> *procedure* LAYOUT − D.D.C. (u, ℓ; c; EX);
> *begin* Connect upper terminal to y = $\sigma(u)$ and lower terminal to
> $\qquad$ y = -$\sigma(\ell)$;
> $\qquad$ r : = net continuing in column c;
> $\qquad$ *if* state (c) ϵ { ⌐, ⌐, ⊣ } *then*
> $\qquad\qquad$ *begin* Q ⇐ $\sigma(r)$;
> $\qquad\qquad\qquad$ connect y = $\sigma(r)$ and y = -$\sigma(r)$;
> $\qquad\quad$ *end*
> $\qquad$ *if* state (c) ϵ { ⊣, ⊣ } *then*
> $\qquad\qquad$ *begin* e : = net continuing at column c;

if $\sigma(r) < \sigma(e)$ *then*

    begin $Q \Leftarrow \sigma(r)$;

        connect $y = \sigma(r)$ and $y = -\sigma(r)$;

    end

    else begin EX: $= $ EX $+ 1$;

        $P \Leftarrow r$;

    end

end

if state(c) $= \; \exists \;$ *then*

  begin EX: $=$ EX $+1$;

    if $\sigma(u) > \sigma(\ell)$ *then*

      begin connect $y = \sigma(\ell)$ and $y = -\sigma(\ell)$;

        $P \Leftarrow u$;   (*extended net*)

        $Q \Leftarrow \sigma(\ell)$;   (*available tracks*)

    end

    else begin connect $y = \sigma(u)$ and $y = -\sigma(u)$;

        $P \Leftarrow \ell$;

        $Q \Leftarrow \sigma(u)$;

        end

  end

end

Lemma 2: Any d.d. column can be processed without increasing the number of tracks.

Proof: When any of column states $\neg$, $\lrcorner$, $\dashv$ is processed, two tracks become free. For column state $\exists$, two tracks become free and one net is extended, and finally, for column states $\perp$ and $\dashv\!\!\!\top$ either two tracks are freed or one net is extended, depending on the relative positions of the tracks carrying the terminating and the continuing nets. So, no tracks are added while a d.d. column is being processed. $\square$

We now consider the handling of d.i. columns. Recall that at a d.i. column one or two nets begin. First suppose that just one net N_i begins (states $\llcorner$, $\ulcorner$, $\vdash$, $\perp$, $\top$). States $\llcorner$, $\ulcorner$, $\vdash$ are handled in a straightforward manner: if there is no extended net, by assigning two unused tracks to N_i; otherwise by connecting strands of an extended net together and running N_i on the tracks previously used by the extended net. There remain states $\perp$ and $\top$, of which we just consider $\perp$. Referring to Figure 9, let N_i be the new net, N_j be the continuing net and N_h be an extended net (if any). If $\sigma(j) > \sigma(i)$, then all the connections are straightforward and N_i is assigned the tracks previously used by N_h (Figure 9a). If $\sigma(j) < \sigma(i)$, then N_j cannot be connected to $y = -\sigma(j)$, so we shall run N_j on $y = -\sigma(i)$ temporarily (in the channel columns where N_i is not lower-active). The two tracks $y = -\sigma(i)$ and $y = -\sigma(j)$ are connected at the closest column e to the right of c with the following property: the lower terminal is labeled either i, or j, or **Λ** (Figure 10). (Note that between columns c and e net N_j has *three* strands.) Finally, when two nets start at c (state(c) = ⊨), if there is just one or no extended net, then the connections to upper (u) and lower (ℓ) terminals are straightforward as in Figures 9a and 9b. If instead there are two or more extended nets, then only one will be connected at column c and the others will be further extended to the right. The integer parameter EX is updated (decreased) as the extended nets are connected. The preceding discussion is formalized in the subroutine LAYOUT − D.I.C. (u, ℓ; c; EX). This subroutine makes use of additional data structures to handle nets temporarily assigned two tracks in either (lower or upper) half-channel. Specifically, for the upper portion of the channel, we shall use two binary search trees

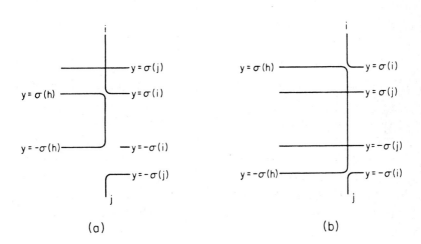

(a)   (b)

Figure 9. Handling of a density-increasing column; N_i is the beginning net, and N_h is an extended net.

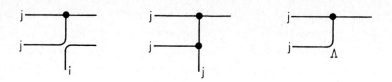

Figure 10. Temporary extension of a net on the inactive track of a beginning net.

AU and AU⁻¹ (and analogously, AL and AL⁻¹ for the lower portion of the channel). Both AU and AU⁻¹ are organized on the basis of the index of a net, and support operations of MIN, MEMBER, INSERT, and DELETE in time logarithmic in their size. Their function is the following: AU(t) = s (i.e., search of AU for member t) means that the third track used by t is $\sigma(s)$; conversely, AU⁻¹(t) = s means that track $\sigma(t)$ is used, as a third track, by s.

procedure LAYOUT − D.I.C. (u, ℓ; c; EX)
f : = 0;
begin if EX ≠ 0 *then*
    begin r ⇐ P;   (* P is the queue of the extended nets*)
        EX : = EX − 1;
        Connect $\sigma(r)$ and -$\sigma(r)$;
        t1 : = $\sigma(r)$;   (* t1 is a track made available*)
        f : = 1;   (* f = 1 means that the column is occupied
                       by a wire between y=$\sigma(r)$ and y=-$\sigma(r)$*)
   end
   else t1 ⇐ Q;   (* Q is the queue of available tracks *)
  if state (c) ϵ { ⌐, L, ⊢ ,⊤̵, ⊥̵ } *then*
    begin if u ≠ **Λ** *and* state(c) ≠ ⊤̵ *then* $\sigma(u)$: = t1;
       if ℓ ≠ **Λ** *and* state(c) ≠ ⊥̵ *then* $\sigma(\ell)$: = t1;
       if state(c) = ⊥̵ *then*
           begin if f = 0 *or* (f = 1 *and* $\sigma(u)$ > t1) *then*
                  t1:= $\sigma(u)$;
             if f = 1 *and* $\sigma(u)$ < t1 *then*
                begin AU (u) : = ℓ;
                      AU⁻¹ (ℓ) : = u
             end

$$end$$

if state (c) = $\perp\!\!\!\top$ *then*

    begin if $f = 0$ *or* $(f=1$ *and* $-\sigma(\ell) < -t1)$ *then*

          $-t1 := -\sigma(\ell)$;

        if $f = 1$ *and* $-\sigma(\ell) > -t1$ *then*

            begin $AL(\ell) := u$;

                   $AL^{-1}(u) := \ell$

          end

    end

Connect upper terminal to t1 with a left bend;

Connect lower terminal to -t1 with a right bend;

  end

else begin $\sigma(u) := t1$;   (*state (c) = $\vdash$ *)

        connect upper terminal to t1 with a left bend;

        $t2 \leftarrow Q$;

        $\sigma(\ell) := t2$;

        if $f = 1$ *and* $t2 < t1$ *then*

          begin $AL(\ell) := u$;

               $AL^{-1}(u) := \ell$;

               $t2 := t1$

         end

        else Connect lower terminal to -t2 with a right bend;

    end

end

Lemma 3: Any d.i. column c can be processed using no more than 2δ tracks.

Proof: It is sufficient to show that at a d.i. column c: (1) no tracks are added if $d_R(c) \leq d_m(c)$ (recall that $d_R(c)$ denotes the local density in $(c, c+1)$), and $d_m(c) = \max_{i<c} d_R(i)$; (2) Two (symmetric) tracks are added if $d_R(c) = d_m(c) + 1$ (this also implies that $d_R(c-1) = d_m(c)$); (3) four tracks are added if $d_R(c) = d_m(c) + 2$ (this also implies that $d_R(c-1) = d_m(c)$).

Indeed, if $d_R(c) = d_m(c) + 1$ or $d_R(c) = d_m(c) + 2$, new (symmetric) tracks are added and the column is laid out in a straightforward manner (as had been described in the handling of d.i. columns). Instead, if $d_R(c) \leq d_m(c)$, then as described earlier, states $\ulcorner$, $\llcorner$, and $\vdash$

are trivially handled by connecting the new net to a free track (or a track occupied to the left of c by an extended net). States $\vdash$, $\perp$ and $\vdash$ are also handled trivially by running the new net(s) on a free track (or tracks occupied to the left of c by an extended net).

The preceding discussion reveals that at a column c with $d_R(c) = \delta$ ($\delta = \max_i d_R(i)$) the number of occupied tracks is 2δ and the number of tracks never exceeds 2δ in any column i for all $i > c$. $\square$

The layout procedure scans the channel column by column from left to right. It calls the appropriate subroutines according to the state of the current column. We shall formalize this in the subroutine LAYOUT1.

procedure LAYOUT1
begin c : = 1;   (* first column *)
        EX : = 0;   (* no extended nets *)
        while EX ≠ 0 *or* there is any d.i.c. left *do*
                begin u : = upper terminal in column c;
                        ℓ : = lower terminal in column c;
                        if AU ≠ ∅ *then*

                begin if $AU^{-1}(u) \neq \Lambda$ *then*
                        begin s : = $AU^{-1}(u)$;   (* third strand of s on $\sigma(u)$ *)
                                connect $y = \sigma(s)$ and $y = \sigma(u)$;
                                connect u to $y = \sigma(u)$ with left bend;
                                delete $AU^{-1}(u)$ and AU(s)
                        end
                        if AU(u) ≠ Λ *then*
                        begin s : = AU(u);   (* third strand of u on $\sigma(s)$ *)
                                connect u to $y = \sigma(s)$ and $y = \sigma(u)$;
                                delete AU(u) and $AU^{-1}(s)$
                        end
                        if u = Λ *then*
                        begin t : = min AU;
                                s : = AU(t);
                                connect $y = \sigma(s)$ to $y = \sigma(t)$;
                                delete AU(t) and $AU^{-1}(s)$
                        end
                end
                if AL ≠ ∅ *then* (* analogous to the above when AU ≠ ∅ *)
                if state(c) = ↑ *then* connect upper and lower terminals;

> *else if* c = d.p.c. *then call* LAYOUT −D.P.C. (u, ℓ; c);
> *else if* c = d.d.c. *then call* LAYOUT−D.D.C. (u, ℓ; c; EX);
> *else if* c = d.i.c. *or* empty *then call* LAYOUT−D.I.C. (u, ℓ;
> c; EX);
> c : = c+1

end

end.

Theorem 1. Any GCRP with essential density δ, can be laid out in 2δ − 1 tracks using the LAYOUT1 algorithm.

Proof: Referring to Lemmas 1, 2 and 3, it can be noted that only at d.i. columns we add tracks. We have also shown in Lemma 3 that the number of occupied tracks is twice the essential density, namely 2δ. A closer look at the algorithm (LAYOUT1) reveals that tracks 1 and -1 are always occupied by the same net (Property 2) so we can merge them. As a result, only 2δ-1 tracks are required to lay out any GCRP with essential density δ. □

2.2   An Improved Algorithm.

The algorithm we have presented in the last subsection was intended to provide intuition for a possible solution to any GCRP and to establish an upper bound to the number of used tracks. The main feature of this algorithm is that, in conjunction with heuristics, it can result in a rather efficient and yet provably good solution to any GCRP. The first thing to note is that the second strand of a net should be added only when it is necessary: indeed, it might very well happen that for some nets we never need to add the second strand. Second, it is only natural to splice the split nets (nets having more than one strand) as soon as it is feasible in a left to right scan. We shall name the improved technique "LAYOUT2". We shall see later that in this method the two strands of a net may be no longer symmetric, so we shall use the notation $\sigma'(t)$ to denote the track used by the second strand of a net t (as opposed to $-\sigma(t)$ in "LAY-OUT1"), with the convention that $\sigma'(t) < \sigma(t)$.

We now look at the handling of the column states by "LAYOUT2". The notation of inclusion between states of a column is defined in a natural way, e.g., ¬ ⊆ ¬L . A column s is said to be a T-column if ⊤ , ⊥ ⊆ state(s) (states ⊥ , ⊤ , ╬ , ╠ , ╬ , ╬ , ╬ are T-states). Handling of any non-T-column is straightforward, as shown in the discussion of "LAYOUT1", as is the handling of states ¬ and ⊥ . So, we need to consider states ╠ , ╬ , ╬ , ╬ and ╬ .

We begin with the two d.i. T- columns (╠ , ╬), of which we just need to consider ╠ (the other case being handled symmetrically). If the net N_u, to be connected to the upper terminal u, has two strands, then we can connect $y = \sigma(u)$ and $y = \sigma'(u)$ and run the beginning net N_ℓ on $y = \sigma'(u)$ as shown in Figure 11a. If instead, N_u has only one strand but there is a free track at $y = f$, with $f > \sigma(u)$, then we can connect $y = \sigma(u)$ to $y = f$ and continue N_u on track f, while N_ℓ is assigned track $\sigma(u)$ (Figure 11b). Other variations are being handled trivially.

Next we shall turn our attention to the handling of the two d.d. T- columns (╬ , ╬) of which we just need to consider ╬ (the other case will be handled symmetrically). The policy is to connect the lower terminal ℓ to $y = \sigma(\ell)$ (and also to $y = \sigma'(\ell)$ if it exists) and connect the upper terminal u to $y = \sigma(u)$ if $\sigma(u) > \sigma(\ell)$ (Figure 12a), or to $y = \sigma(\ell)$ with a left bend, otherwise (Figure 12b). Other variations are being handled trivially.

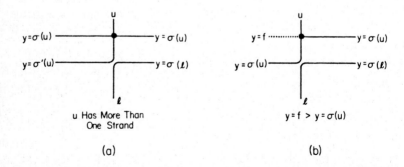

Figure 11. Handling of a d.i. T-column

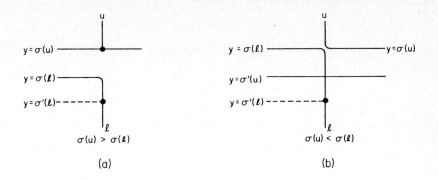

Figure 12. Handling of a d.d. T-column

It remains to consider the handling of a d.p. T-column ($\dashv\vdash$). Denoting, as usual, by N_u and N_ℓ the two nets having the upper and lower terminals, respectively, there are eight possible cases (ignoring trivial symmetries), depending upon the number of strands of N_u and N_ℓ and their relative positions. The handling of these cases is illustrated in Figure 13, and deserves no further comment.

The preceding improvements yield the following result:

Theorem 2: Any GCRP with essential density δ can be routed in $\delta + \alpha$ tracks for $0 \leq \alpha \leq \delta - 1$ by the "LAYOUT2" algorithm.

While δ is a lower bound for the minimum number of horizontal tracks, we have shown an upper bound of $2\delta - 1$ for the number of tracks required by any GCRP using the "LAYOUT1" algorithm. By saving extra tracks (beyond δ) in the "LAYOUT2" algorithm we frequently can solve any GCRP in fewer than $2\delta - 1$ tracks. Simulations reveal that α, the number of tracks used beyond δ, is rather small. In fact for some GCRP α is equal to zero.

Next we shall turn our attention to a special case of GCRP, namely a two-terminal net CRP. We shall route a CRP by the "LAYOUT2" algorithm. Of the twenty possible column states, only the following ten states ($|$, $\uparrow$, $\neg$, $\lrcorner$, $\dashv$, $\llcorner$, $\ulcorner$, $\vdash$, $\dashv\llcorner$, $\dashv\ulcorner$) may arise.

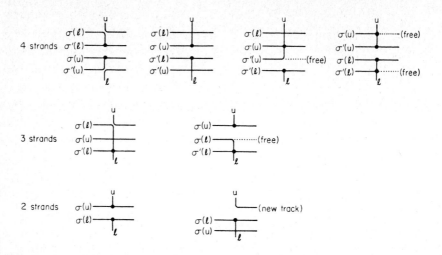

Figure 13. Handling of a d.p. T-column

Recall that the algorithm scans the channel from left to right and pro-
cesses successively all the columns. Note that Property 1 holds in all
columns   except when state(c) $= \dashv$. The processing of this column
is straightforward if $\sigma(u) > \sigma(\ell)$ (where, as usual, u is the upper terminal and ℓ the lower terminal) as in Figure 14a. If, on the other hand,
$\sigma(u) < \sigma(\ell)$, then we shall connect the upper terminal u to $y = \sigma(u)$
and connect the lower terminal to $y = \sigma'(\ell)$ where we have assigned
$\sigma'(\ell) := \sigma(u)$ as in Figure 14b.

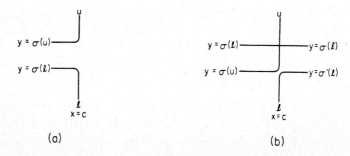

Figure 14. Layout cases for the two-terminal net CRP.

We shall make use of a priority queue **P** (FIFO) of the extended nets. So N_ϱ is introduced into the queue (Figure 14b); but in the interval (c, e), where e is the closest density-increasing or empty column (states ⌐ , ⌐, ⌐, ⊨) to the right of c, the number of occupied tracks will not exceed the local density $d_L(c)$-2 $\leq$ δ-2, so that we can connect the two strands of the extended net N_ϱ at column e (Figure 15).

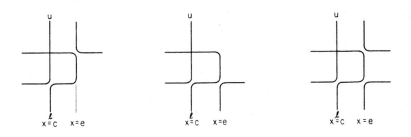

Figure 15. Layout of extended nets in the two-terminal net CRP.

The preceding discussion gives us the following result:

Theorem 3. Any two terminal net CRP with density δ can be routed using δ tracks by the "LAYOUT2" algorithm.

3. Three-layer wirability

We now return to the general problem and show that any layout produced by "LAYOUT2" can be wired with three layers without increasing the number of tracks used in the layout phase.

Our arguments are heavily based on the wirability theory developed by Preparata and Lipski [PL], to which the reader is referred. We begin by observing that in the layout of multiterminal nets there is one more type of grid points, the "T", in addition to the five standard ones encountered in the layout of two-terminal nets (the "crossing", the "knock-knee", the "bend", the "straightwire", the "empty"); these six types are illustrated in Figure 16. Following the arguments in [PL], suppose that in a given layout W we replace each non-knock-knee grid-point by a crossing: it is very simple to show that if the resulting layout W* —— a "full" layout —— can be wired with three layers, so can the original W.

Note that in the transformation from W to W* we have, in essence, obtained a two-terminal net problem. We must only verify that it satisfies the three-layer-wirability sufficient condition proposed by Preparata and Lipski:

We begin by observing the following facts:

(1) at a d.p. column (states ⌐L and ⌐Γ) we have at most one knock-knee.

(2) at a d.d. column (states ⊥ , ⊤ and ⊣) we have at most one knock-knee.

(3) at a d.i. column we have either one knock-knee (states ⌐, L, ⊢, ⊤) or two (state ⊢ , and ⊢) if the priority queue of the extended nets is not empty.

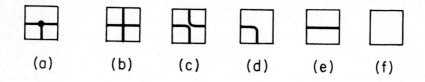

(a)   (b)   (c)   (d)   (e)   (f)

Figure 16.   Types of grid-point in multiterminal net layout:
(a): T, (b): crossing, (c): knock-knee; (d): bend;
(e): straightwire; (f): empty.

We can therefore assume that every column contains one of the following: (i) a single knock-knee of the form ⌐Γ , (ii) a single knock-knee of the form ⌐L ,or (iii) two knock-knees, of which the lower one is of the form ⌐Γ and the upper one is of the form ⌐L . As is customary, we graphically denote a knock-knee grid point by means of a $\sqrt{2}$ - length diagonal, centered on the grid point and crossing both wires as shown in Figure 17.

Figure 17. Graphical equivalents (diagonals) of knock-knees.

With this convention we can convert any layout to a "diagonal diagram". We augment the diagram by adding dummy tracks 0 and δ +α +1, placing ▧ (knock-knee of the form ⌐L) on track δ +α +1 for every column of type (i), and placing ◹ (knock-knee of the form ⌐L) on track 0 for every column of type (ii). We shall refer to these additional diagonals as *dummy* diagonals. In this way we can restrict ourselves to the case where all nonempty columns are of type (iii).

Preparata and Lipski have shown that any diagonal diagram of this type corresponds to a full layout that is wireable with three layers, provided that we allow for layout modifications in correspondence with special pattern of diagonals. We now show that, in the layouts produced by our algorithms, such patterns never arise.

A representative of the special diagonal pattern is shown in Figure 18a, together with the corresponding fragment of layout. This layout shows that two disconnected strands of a net have been spliced in column c+1 and state (c) is either ⌐L (Figure 18a) or ⊨ (Figure 18c). In either case, our policy to splice a two-strand net as soon as possible, will generate the layouts shown in Figures 18b, d (i.e., one extended net is spliced at column c rather than c+1), so that the cases shown in Figures 18a, c never arise.

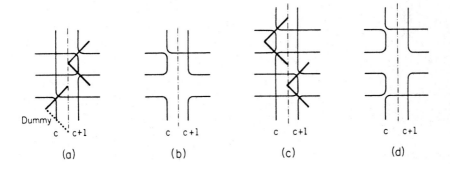

Figure 18.

Thus, as an immediate consequence of the above observations and of the results established in [PL], we conclude:

Theorem 4. Any layout produced by the "LAYOUT2" algorithm that uses $\delta + \alpha$ tracks (for $0 \leq \alpha \leq \delta - 1$) can be wired in three layers using $\delta + \alpha$ tracks.

Since both the layout and the wiring algorithms scan through the channel column-by-column, they could run simultaneously.

To illustrate the method, we now give two examples: a GCRP (Figure 19), where $\delta = 5$ and $\alpha = 1$, and a two-terminal net CRP (Figure 20), with $\delta = 8$ and $\alpha = 0$.

Before closing the section, we briefly analyze the time performance of the proposed algorithm LAYOUT2. Denoting by C the number of columns, the input is assumed to be in the form of a sequence of pairs $((u_i, \ell_i) \mid i = 1,..., C)$ where $u_i, \ell_i \in \{1,.., n\}$ specify the upper and lower terminals of column i; these pairs are stored in an array. Processing of each column i involves some actions on the data structure required by the procedures described earlier (priority queues Q and P, search trees AU, AU^{-1}, AL, AL^{-1}, and the array σ). Inspection of the algorithms reveals that the number of operations performed at each column is bounded by a constant; in addition, each of these operations takes time at most logarithmic in the size of the data structure (which is O(n)). Thus we conclude:

Theorem 5. "LAYOUT2" runs in time proportional to Clog n, where C is the number of columns and n is the number of nets.

It is relatively straightforward to show that the wiring task can also be accomplished within the same time bound.

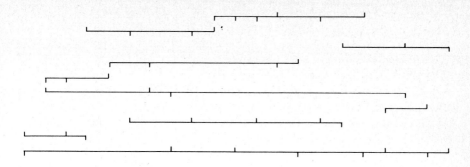

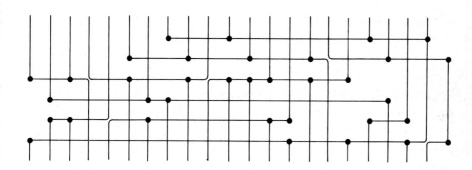

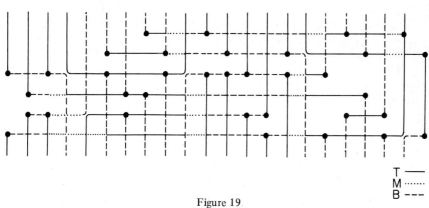

T ———
M ·······
B – – –

Figure 19

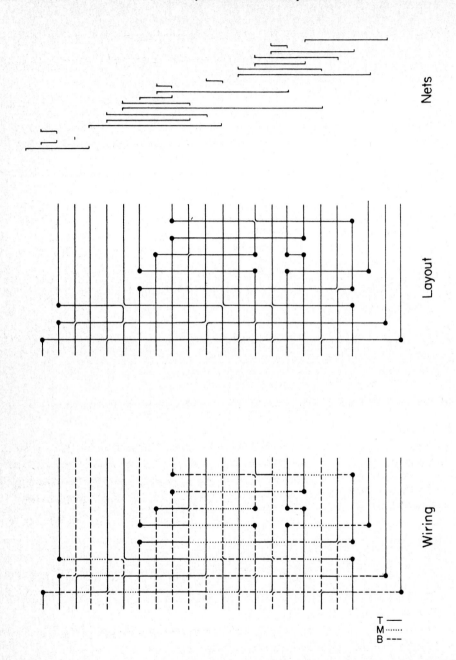

Figure 20

References

[B] Baker, B. S., private communication

[BB] Bolognesi, T. and D. J. Brown, "A channel routing algorithm with bounded wire lenght," draft.

[BP] Brown, D. J. and F. P. Preparata, "Three-layer channel routing of multiterminal nets" draft.

[L] Leighton, F. T., "New lower bounds for channel routing," draft, 1981.

[PL] Preparata, F. P. and W. Lipski, Jr., "Optimal three-layer channel routing," *IEEE Transactions on Computers, C-33, 5,* May 1984, pp. 427-437.

[RBM] Rivest, R. L., A. Baratz and G. Miller, "Provably good channel routing algorithms," *Proc. 1981 CMU Conference on VLSI Systems and Computations,* Oct. 1981, pp. 153-159.

[RF] Rivest, R. L. and M. Fiduccia, "A greedy channel router." *Proceedings 19th IEEE Design Automation Conference,* 1982, pp. 418-424.

Annals of Discrete Mathematics 25 (1985) 281-290
© Elsevier Science Publishers B.V. (North-Holland)

CONSISTENCY OF QUADRATIC BOOLEAN EQUATIONS AND THE KÖNIG–EGERVÁRY PROPERTY FOR GRAPHS(°)

Bruno SIMEONE

Department of Statistics and Probability

University of Rome "La Sapienza"

The purpose of the present note is to point out the close connection between the consistency of quadratic boolean equations and the maximum matching-minimum covering equality in graphs. Some of the results can be extended to arbitrary boolean equations and to hypergraphs.

We shall denote by latin letters x, y, ... boolean variables; $\bar{x}, \bar{y}, \ldots$ will denote their complements, while greek letters $\xi, \eta \ldots$ will denote *literals* i.e. complemented or uncomplemented variables. A *quadratic boolean equation* has the form

$$T_1 \vee \ldots \vee T_m = 0, \tag{1}$$

where each *term* T_i is the conjunction of at most two literals. The l.h.s. of (1) is called a quadratic boolean form.

Let G = (V, E) be a graph. A *matching* of G is a set of edges such that no two of them have a common vertex. A (vertex) *covering* (or

(°) The present work is part of a doctoral thesis written at the Department of Combinatorics and Optimization, University of Waterloo, under the supervision of Prof. P.L. Hammer.

transversal) of G is a subset C of vertices such that every edge has at least one endpoint in C. Finally, an *independent set* of G is a subset I of vertices no two of which are adjacent.

The maximum cardinality of a matching will be denoted, as customary, by $\mu(G)$; the minimum cardinality of a covering by $\beta(G)$. It is well known and easy to see that

$$\mu(G) \leq \beta(G) \tag{2}$$

A graph is said to have the *König-Egerváry* (KE) *property* if equality holds in (2).

Given a matching M, a vertex v of G is said to be *matched* or *free* according as v is the endpoint of an edge in M or not.

A pair (S, M), where $S \subseteq V$ and M is a matching, will be called a *rake* if every vertex in S is the endpoint of exactly one edge in M.

Lemma 1. (Klee, as reported in [4, p. 191]) *A graph has the König-Egerváry property if and only if there exist a (necessarily minimum) covering C and a (necessarily maximum) matching M such that (C, M) is a rake.*

Lemma 2. (Gavril [3]). *A graph has the König-Egerváry property if and only if, for every maximum matching M and every minimum covering C, the pair (C, M) is a rake.*

Let G be an arbitrary graph and let $M = \left\{ e_1, \ldots, e_q \right\}$ a maximum matching of G. The set F of free vertices has $p = |V| - 2q$ elements. Let us associate with G and M a quadratic boolean form $\Phi(x; G, M)$ in $n = p + q$ variables $x_1, \ldots, x_n$ in the following way.

For each $e_i \in M$, we associate the literal x_i with one of the endpoints of e_i and the literal x_i with the other endpoint. Further, we associate a literal x_i, $i = q + 1, \ldots, n$, with each free vertex. Finally, denoting by $\xi(v)$ the literal associated with vertex v, we set

$$\Phi(x; G, M) = (\underset{uv \in E-M}{V} \xi(u) \xi(v)) V (\underset{w \in F}{V} \overline{\xi(w)}) \qquad (3)$$

The quadratic boolean equation ⌄ $(x; G, M) = 0$ is called *the boolean equation associated with* G *and* M. Strictly speaking such an equation is not uniquely determined by G and M, because of the freedom in numbering the edges of M and the free vertices of G, and in associating either x_i or $\overline{x}_i$ with an endpoint of e_i. However, any two equations generated in this way are equivalent, in the sense that one can easily build a bijection between their solution sets.

As an example, the boolean equation associated with the graph and the matching in Fig. 1 (edges in the matching are represented by thick lines)

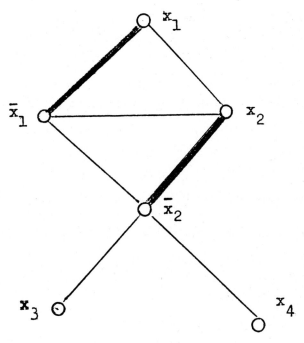

Fig. 1

B. Simeone

is

$$x_1 x_2 \ \lor \ \bar{x}_1 x_2 \ \lor \ \bar{x}_1 \bar{x}_2 \ \lor \ \bar{x}_2 x_3 \ \lor \ \bar{x}_2 x_4 \ \lor \ \bar{x}_3 \ \lor \ \bar{x}_4 \ = 0 \qquad (4)$$

Theorem 3. *Let* G *be a graph and* M *an arbitrary maximum matching of* G.

Then G *has the König-Egerváry property if and only if the boolean equation* $\Phi(x; G, M) = 0$ *is consistent.*

Proof. Assume that the boolean equation is consistent and let x^* be a solution vector. Let I be the set of all vertices $v \in V$ such that the associated literal $\xi(v)$ takes the value 1 in the given solution x^*. The set I must be independent; hence $C = V - I$ must be a covering. Moreover, all vertices in C must be matched, because $F \subseteq I$. On the other hand, every edge of M must have exactly one endpoint in C and one in I. Hence (C, M) is a rake. Then, by the "if" part of Lemma 1, G has the KE property. Conversely, assume that the KE property holds for G. If C is an arbitrary minimum covering of G, then (C, M) must be a rake by the "only if" part of Lemma 2.

The set $I = V - C$ is independent and must include F, because all nodes in C are matched. Hence, if we assign the value 0 or 1 to $\xi(v)$ according to whether $v \in C$ or $v \in I$, we get a solution of the equation $\Phi(x; G, M) = 0$. ∎

For example, for the graph G and the maximum matching M in Fig. 1, the associated boolean equation (4) has no solution. On the other hand, one has $\mu(G) = 2$ and $\beta(G) = 3$.

Conversely, given a quadratic boolean equation, one can build a graph in such a way that the equation is consistent if and only if the graph has the König-Egerváry property.

A more general construction can be given for arbitrary boolean equations: in this case, a hypergraph, rather than a graph, is associated with the boolean equation.

Let $H = (V, \mathscr{E})$ be a hypergraph. We assume that every edge has at least two elements. A *matching* M is a set of edges any pair of which has empty intersection. A (node) *covering* (or *transversal*) is a set C of vertices which has non-empty intersection with every edge. An *independent set* is a set I of vertices which does not contain any edge. As for the case of graphs, a set of vertices C is a covering if and only if its complement $I = V-C$ is an independent set.

If $\mu(H)$ and $\beta(H)$ are defined as for graphs, the relation

$$\mu(H) \leq \beta(H) \tag{5}$$

holds also for hypergraphs. A hypergraph is said to have the *König-Egerváry property* if equality holds in (5). Consider now an arbitrary boolean equation $\psi(x) = 0$ in n variables $x_1, \ldots, x_n$ and with m terms. Without loss of generality, we can assume that:

a) the equation has no linear term ξ. For otherwise, ξ must take the value 0 in every solution (if any), and we are led to a reduced equation in $n-1$ variables, which is consistent if and only if the original one is such.

b) All literals $x_1, \ldots, x_n; \bar{x}_1, \ldots, \bar{x}_n$ explicitly appear in the equation. For, if the literal ξ is not present, we can assign the value 0 to $\bar{\xi}$ and again we get a reduced equation in $n-1$ variables which is consistent if and only if the original one is such.

Let us associate with the boolean form ψ a hypergraph $H_\psi = (V_\psi, \mathcal{E}_\psi)$ as follows. The vertices of H_ψ are the 2n literals $x_1 \ldots,$ $x_n; \bar{x}_1, \ldots, \bar{x}_n$.

The edges of H_ψ are the terms of ψ (a term is thought here as of a set of literals); in addition, H_ψ has the n edges $x_i\bar{x}_i$ $i = 1, \ldots, n$. H_ψ is called the *matched hypergraph* of ψ. We note that, because of (a), each edge has at least two elements and that, because of (b), each vertex is contained in some edge.

Theorem 4. *A boolean equation* $\psi(x) = 0$ *is consistent if and only if its matched hypergraph* H_ψ *has the König-Egerváry property.*

Proof. Let x* be a solution of the equation $\psi = 0$, and let us define two sets of vertices of H_ψ as follows:

$$C \equiv \left\{ x_i : x_i^* = 0 \right\} \cup \left\{ \bar{x}_i : x_i^* = 1 \right\} \text{ and } I \equiv \left\{ x_i : x_i^* = 1 \right\} \cup \left\{ \bar{x}_i : x_i^* = 0 \right\}$$

One has $|I| + |C| = |V_\psi| = 2n$ and $|I| = |C|$. Hence $|C| = n$. Since x* is a solution of equation $\psi(x) = 0$, C is a covering of H_ψ. On the other hand, since every edge has cardinality at least two, one must have $\mu(H_\psi) \leq n$. Hence the set of edges $M \equiv \left\{ x_i \bar{x}_i : i = 1, \ldots, n \right\}$ is a maximum matching. In view of (5), one has $n = |M| = \mu(H_\psi) \leq \beta(H_\psi) \leq |C| = n$. Thus $\mu(H_\psi) = \beta(H_\psi)$. Conversely, if H_ψ has the KE property, then $\beta(H_\psi) = \mu(H_\psi) = |M| = n$. Therefore, if C is a minimum covering, every edge of M must contain exactly one element of C. Thus, for all i, either x_i or $\bar{x}_i$, but not both, belongs to C. Define a binary n-vector x* by

$$x_i^* = 0 \text{ if and only if } x_i \in C \qquad (i=1, \ldots, n)$$

Since C is a covering of H_ψ the vector x* is a solution of the equation $\psi(x) = 0$. ∎

In the particular case when the boolean equation $\psi = 0$ is quadratic, the matched hypergraph is actually a graph, the *matched graph* G_ψ of ψ.

Example. Consider the quadratic boolean equation

$$\psi(x) \equiv x_1 \bar{x}_2 \vee x_1 x_3 \vee \bar{x}_1 x_5 \vee x_2 \bar{x}_4 \vee \bar{x}_3 \bar{x}_5 \vee x_4 x_5 = 0 \qquad (6)$$

The matched graph G_ψ is shown in Fig. 2.

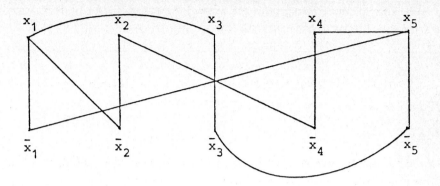

Fig. 2

For such graph one has $\mu(G_\psi) = \beta(G_\psi) = 5$. On the other hand, equation (6) admits the solution $x^* = (0, 1, 1, 1, 0)$.

A combinatorial characterization of graphs with the König-Egerváry property has been given by Deming [1] and Sterboul [5]. Using their results, one immediately obtains a characterization of consistent quadratic boolean equations in terms of the absence of certain "obstructions" in G_ψ.

Given a graph G and a matching M of G, an *alternating path* (with respect to M) is any path whose edges are alternately in M and not in M.

A *blossom* (with respect to M) is an odd cycle $C = v_0 v_1 \ldots v_{2q} v_0$ such that the edges $v_0 v_1$ and $v_{2q} v_0$ do not belong to M and $v_0 v_1 \ldots v_{2q}$ is an alternating path.

The vertex v_0 is called the *tip* of the blossom: v_0 is the only vertex of C which is not incident to any edge of C belonging to M.

A *posy* (with respect to M) is a pair of blossoms C_1, C_2 together with an odd alternating path **P** connecting the tips of C_1 and C_2 and such that both the first and the last edge of **P** belong to **M**.

Corollary 5. *A quadratic boolean equation* $\psi(x) = 0$ *is consistent if and only if its matched graph* G_ψ *has no posy with respect to the matching* $M = \left\{ x_i \bar{x}_i : i = 1, \ldots, n \right\}.$

Proof. A result of Deming [1] and of Sterboul [5] implies that a graph G with a perfect matching M has the König-Egerváry property if and only if G has no posy with respect to M. Then the thesis follows from Theorem 4. ∎

By Theorem 4, the problem of checking the consistency of a quadratic boolean equation $\psi(x) = 0$ is reducible to the problem of recognizing whether the matched graph G_ψ has the König-Egerváry property. The latter problem can be solved by an algorithm proposed by Gavril [3].

Actually, the special structure of G_ψ allows some simplifications:

(i)   it is not necessary to find a maximum matching of G_ψ, as required in general by Gavril's algorithm, because the perfect matching $M = \left\{ x_i \bar{x}_i : i = 1, \ldots, n \right\}$ is at hand;

(ii) G_ψ has no free vertices with respect to M.

The time complexity of the algorithm is $O(m+n)$, where n is the number of variables and m is the number of terms.

For further details, one may consult Gavril's paper [3]. It should be mentioned that the resulting algorithm turns out to be very similar to an algorithm for 2-satisfiability sketched in [2].

The case when G_ψ is bipartite deserves further attention. A quadratic boolean equation is called *mixed* if each of its terms has the form $x_i \bar{x}_j$.

Given a boolean equation (not necessarily quadratic) a *switch* on the variable x is a transformation of variables $x = \bar{y}$: every occurrence of x in the equation is then replaced by $\bar{y}$ and every occurrence of $\bar{x}$

by y. Similarly one defines a switch on a set S of variables.

Theorem 6. *For an arbitrary quadratic boolean equation $\psi = 0$ the following propositions are equivalent:*

1) *The boolean equation $\psi = 0$ is consistent, and whenever x is a solution, also $\bar{x}$ is a solution.*

2) *By a switch on some set of variables, the boolean equation $\psi = 0$ can be trasformed into an (equivalent) mixed boolean equation.*

3) *The matched graph G_ψ associated with ψ is bipartite*

Proof. 1) $\Rightarrow$ 3).

If x^* is a solution of the equation $\psi = 0$, the set $I = \left\{ x_i : x_i^* = 1 \right\}$ $\cup \left\{ \bar{x}_i : x_i^* = 0 \right\}$ is independent in G_ψ. Since $\bar{x}^*$ is also a solution, the set $J \equiv \left\{ x_i : x_i^* = 0 \right\} \cup \left\{ \bar{x}_i : x_i^* = 1 \right\}$ is also independent. But I and J form a partition of the vertex set of G_ψ. Hence G_ψ is bipartite.

3) $\Rightarrow$ 2)

If G_ψ is bipartite, its vertex set V can be partitioned into two independent sets I and J. Both I ($=V-J$) and J ($=V-I$) are also coverings. Hence, in each term of ψ, exactly one literal belongs to I and the other one to J. Moreover, both (I, M) and (J, M) must be rakes. Hence the set $S \equiv \left\{ i : \bar{x}_i \in I \right\}$ must coincide with the set $\left\{ i : x_i \in J \right\}$. It follows that the switch on S transforms all literals in I into uncomplemented variables, and all literals in J into complemented ones. Hence, after the switch on S all the terms of the transformed equation are of the form $x_i \bar{x}_j$.

2) $\Rightarrow$ 1)

Property 1) certainly holds for any boolean equation which is mixed. Thus it must hold also for any boolean equation which can be trasformed into a mixed one by a switch on some set of variables. ∎

References

[1] K. W. Deming: *"Independence numbers of graphs-an extension of the König-Egerváry theorem"*, Discr. Math. 27(1979) 23-34.

[2] S. Even, A. Itai, A. Shamir: *"On the complexity of time-table and multi-commodity flow problems"*, SIAM J. on Computing 5(1976) 691-703.

[3] F. Gavril: *"Testing for equality between maximum matching and minimum node covering"*, Inform. Process. Lett. 6(1977) 199-202.

[4] E. Lawler: *Combinatorial Optimization: Networks and Matroids* (Holt, Rinehart and Winston, New York, 1976).

[5] F. Sterboul: *"A characterization of the graphs in which the transversal number equals the matching number"*, J. of Comb. Theory, Ser. B, 27(1979) 228-229.

Annals of Discrete Mathematics 25 (1985) 291-310
© Elsevier Science Publishers B.V. (North-Holland)

ON SOME RELATIONSHIPS BETWEEN COMBINATORICS AND PROBABILISTIC ANALYSIS

M. TALAMO

IASI – CNR, Viale Manzoni 30, Roma

A. MARCHETTI–SPACCAMELA

Dipartimento di Informatica e Sistemistica Università di Roma La Sapienza, Via Eudossiana 18, Roma

M. PROTASI

Dipartimento di Matematica Università dell'Aquila, Via Roma 33, L'Aquila.

1. Introduction

In the past years several polynomial or NP-complete optimization problems have been studied, from a probabilistic point of view, with the aim of finding fast algorithms that achieve for "most cases" an optimal solution or a very good approximate one. To formulate what "most cases" means it is necessary to introduce a probability distribution over the set of problem instances of each size that allows to define a random instance of the problem. Given a random instance of size n it is possible to evaluate z_n^*, the random variable that expresses the value of the optimal solution; furthermore if we consider an algorithm A we can study z_n^A, the random variable that indicates the value of the solution given by the algorithm A.

The starting point of this research has been the observation that several problems on graphs and sets (for instance Max clique [3], Max set-packing [8], Max tree [6]), share common properties from a probabilistic point of view. For example, for a fixed probability distribution (the constant density model) a greedy algorithm, applied to the above problems, gives a solution whose value is, with high probability, equal to the half of the optimal solution. With respect to this observation we want to investigate the combinatorial characteristics which determine this behaviour and provide a nontrivial generalization of the analysis developed for the above mentioned problems.

More specifically we introduce a class of combinatorial optimization problems, characterized by a weak hereditary property on the set of the feasible solutions; that is, given a feasible solution S of size i, there always exists an element a_j in S such that S- $\{a_j\}$ is a feasible solution of size i-1. These problems are called weakly hereditary optimization problems and share some simple combinatorial properties that determine the probabilistic behaviour of the whole class with respect to the greedy algorithm. Note that in this way we find again the results known in the literature for the above problems and we can analyze problems not yet studied.

A second important point concerns the probability distribution which has to be introduced in order to perform the analysis. Generally the results presented in the literature are stated with respect to a fixed probability distribution. In our case, the results are sufficiently powerful to hold for a wide class of probability models, which includes many known distributions. These distribution independent results are particularly interesting because the main objection moved to the probabilistic approach is that we seldom know which is the real distribution of the input that we meet in practical cases.

The paper is organized as follows: in par. 2 we define the class of the weakly hereditary optimization problems and list some important problems belonging to this class. In par. 3 we introduce the probabilistic models that we will use and in par. 4 we study the main combinatorial properties of the class that will allow to perform the analysis. Finally in paragraphs 5 and 6 we study and compare the optimal solution and the solution given by a greedy algorithm.

2. The class of weakly hereditary optimization problems

In this paragraph we present the class of problems that we want to study and we begin to introduce the mathematical tools that we will need in the following. The class that we will study is already known in the literature [5] as the class of the accessible set system (*).

Definition 1. A *maximization problem* P over an alphabet Σ is a pair <INPUT, SOL> where

i) INPUT is the set of instances. Every instance contains a set of 'elements'.

ii) SOL : INPUT → 2^{INPUT} is a mapping that to every instance I ϵ INPUT associates a set of subsets of I representing the feasible solutions;

iii) the optimal solution of I is $z^*(I) = \max_{S\epsilon SOL(I)} |S|$.

Definition 2. A weakly hereditary maximization problem (shortly WHOP) P = <INPUT, SOL>
is a maximization problem such that for every I ϵ INPUT we have

i) $\emptyset$ ϵ SOL (I);

ii) if S ϵ SOL(I) and S ≠ $\emptyset$, then there exists a_j ϵ S such that $S - \{a_j\}$ ϵ SOL(I);

iii) the number of feasible solutions that it is possible to build from a solution of size i depends only on i.

In literature other classes of problems based on hereditary properties have been introduced; in particular we recall the definition of hereditary problem given by Yannakakis [10] of greedoid by Korte e Lovasz [5] and, finally, of random matroid by Reif e Spirakis [7]. We do not formally study the differences among all these definitions, but we limit to observe that it is easy to see that our definition is the weakest one and allows to include all the problems that

---

(*) However we will add a condition of uniformity and since we are interested in evaluating the performance of approximate algorithms, we will directly define the class in an optimization version.

belong to any other class.

Many problems on graphs and sets belong to our class as: MAX-CLIQUE: INPUT is the set of all graphs; given a graph G, SOL(G) is the set of all complete subgraphs (cliques) of G. The condition of weak hereditarity is trivially satisfied.

K − MAX-HYPERCLIQUE: Extension of the max-clique problem to the K-hypergraphs.

MAX-TREE: INPUT is the set of all graphs; given a graph G, SOL(G) is the set of all trees; given a tree T of n nodes, there exists a node i in T such that T-$\{i\}$ is a tree.

MAXIMUM-MATCHING (and related problems such as 3 − DIMENSIONAL MAXIMAL MATCHING): INPUT is the set of all graphs; SOL(G) is the set of all matchings; also in this case the property is trivially verified.

MAX-INDUCED PATH: INPUT is the set of all graphs; SOL(G) is the set of all induced paths of G; given a path, deleting the first or the last node, we will obtain a path.

K − MAX − SET PACKING: INPUT is the set of all finite families of sets; given a family $F = \{S_i\}$ of sets, with $|S_i| = K$, for every i, SOL(F) is the set of all subfamilies formed by disjoint sets S_i; again the property of weak hereditarity is satisfied.

Examples of other weakly hereditary optimization problems (see [2] for definitions) are:

MAX COMPLETE BIPARTITE SUBGRAPH

LONGEST COMMON SUBSEQUENCE

LONGEST SIMPLE PATH

MAX PARTITIONABLE SUBGRAPH INTO TRIANGLES

MAX K−ACYCLIC SUBHYPERGRAPH

3. The probabilistic model and its properties

Since we want to perform our analysis independently, as more as possible, from the choice of a particular probabilistic distribution on the set of the instances, we will define a class M of models which, as we will see, is very general.

Definition 3. Let P be a WHOP. Let I be an instance of P. M is the class of models for which the probability $f(i)$

$$f(i) = \text{Prob} \left\{ S \cup \left\{ a_j \right\} \, \epsilon \, \text{SOL}(I) \, / \, S \epsilon \text{SOL}(I), |S| = i, a_j \epsilon I, a_j \quad S \right\}$$

has to verify the following conditions

There exists a function $d(n) \geqslant 1$ such that

i) $f(i \cdot j) \leqslant (f(i))^{j^{d(n)}}$ for any i, i·j integers

ii) $f(i \cdot j) > (f(i)^j)^{j^{1-\epsilon}}$ for every $\epsilon > 0$

In particular $d(n)$ can be also a constant and in the following for simplicity of notation we will always write d instead of $d(n)$. Intuitively the function $f(i)$ expresses the probability to add one element to a feasible solution of size i obtaining a solution of size i+1.

The requirement that S is a solution is clearly linked to the weak hereditarity property that assures that every solution of size i +1 can be obtained from a solution of size i by adding a new element.

Finally we note that $f(i)$ may be also depending on other parameters such as, for example, the size of the instance I.

Definition 4. Let P be a WHOP. Let I be an instance of size n of P. The expected value $E_n(i)$ of the number of feasible solutions of size i of I is defined in the following way

$$E_n(i) = x_{n,i} \prod_{k=0}^{i-1} f(k)$$

The class of models captured in definitions 3 and 4 allows to include a large number of distributions. All the models more known in the literature in the case of graph problems verify our definitions. Furthermore models in which the probability p_{ij} of existence of an edge (i, j) changes with i and j or is depending on the size n of the instance may be also considered, of course if the conditions on $f(i)$ hold.
We now give particular examples.

For the problems already defined, assuming the constant density model we have the following values of $f(i)$:

Let p be the probability of existence of an edge

1)   Max clique $f(i) = p^i$

2)   K max hyperclique $f(i) = p^{\binom{i}{K-1}}$

3)   Max tree $f(i) = ipq^{i-1}$

4)   Maximum matching $f(i) = p$

5)   Maximum induced path $f(i) = 2pq^{i-1}$

1)   Max clique $f(i) = p^i$

Given a random graph $G_{n,p}$, a set S of i nodes that induces a complete subgraph and a node $a_j \notin S$ we have that the subgraph induced by $\{a_j\} \cup S$ is a clique if and only if a_j is connected to all nodes of S; hence $f(i) = p^i$.

2)   K max hyperclique

By a straightforward generalization of the above argument we have $f(i) = p^{\binom{i}{K-1}}$.

3)   Max tree

Given a random graph $G_{n,p}$, a set S of i nodes that induces a tree and a node a_j not in the tree, we have that the subgraph induced by

$\{a_j\} \cup S$ is a tree if and only if a_j is connected to one and only one node of S. This implies that $f(i) = ipq^{i-1}$ if $q = 1-p$

4) Maximum matching

Given a random graph $G_{n,p}$, and a set S of i nodes that induces a matching, and a couple of nodes a_j such that $S \cap \{a_j\} = \emptyset$, we have that $S \cup \{a_j\}$ is a matching if and only if a_j is an edge. This implies $f(i)=p$.

5) It is easy to see that $f(i) = 2pq^{i-1}$ by an analogous argument to that given in 3).

Assuming a probabilistic model on sets similar to the constant density model it is easy to see that for K-max-set packing $f(i) = p^i$.

4. Relationships between combinatorial and probabilistic properties

In this paragraph, we relate the combinatorial structure of weakly hereditary optimization problems to the expected value $E_n(i)$ of the number of feasible solutions of cardinality i over the instances of size n. The bounds determined for $E_n(i)$ will be used in the following paragraphs for evaluating the value of the optimal and the approximate solution.

In particular we assume that, for any i elements, the number of different solutions we may find is bounded by i^i. Hence, given an instance of cardinality n, the number $\Psi_{n,i}$ of feasible solutions of cardinality i is bounded by $\binom{n}{i} i^i$. We note that this upper bound on the number of solutions is verified by many classical problems on graphs and sets, besides the problems considered in par. 2.

Furthermore we note that a trivial lower bound on the number of feasible solutions is given by $\binom{n}{i}$ as we assume that for any given i objects there always exists a solution.

Now we want to give an approximation of

$$E_n(i) = \chi_{n,i} \prod_{k=0}^{i-1} f(k).$$

Infact, we do not know the value of $\chi_{n,i}$ and, generally speaking, it is impossible to evaluate exactly $\chi_{n,i}$, because of the richness of the combinatorial structure. Observe that when we add an object a to a feasible solution we can obtain different solutions with the same set of objects. For example, let us consider the problem Max-tree and a tree given in fig. 1a. By adding a new node a we can obtain different trees as shown in fig. 1b.

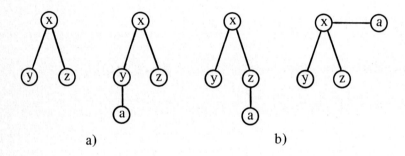

a)   b)

Fig 1. Different solutions obtained by adding a new node

In general this observation implies that we can write f(i) as follows:

$$f(i) = \theta_i \bar{p}_i$$

where $\bar{p}_i$ is the probability that given a solution of i objects, adding a new object we obtain exactly one of the θ_i possible solutions of size $i+1$. For the problems that we will study, $1 \leqslant \theta_i \leqslant i$ and in the following we will do this very general assumption.

For example for the Max-tree problem we have $\theta_i = i$ and $\bar{p}_i = pq^{i-1}$.

Note that $\displaystyle\prod_{k=0}^{i-1} \bar{p}_k$ gives the probability of obtaining a particular feasible solution of i given objects. Therefore, we can write $\Psi_{n,i}$ as follows:

$$\Psi_{n,i} = x_{n,i} \prod_{k=0}^{i-1} \theta_k.$$

Now we are able to give a precise approximation of $x_{n,i}$

Theorem 1. $\binom{n}{i} \leqslant x_{n,i} \leqslant \binom{n}{i} e^i$.

Proof. The first bound $\binom{n}{i} \leqslant x_{n,i}$ can be derived by observing that there exists a probability distribution f' such that $E_n(i) \geqslant \binom{n}{i}$. This implies that $\binom{n}{i} \leqslant E_n(i) = x_{n,i} \prod_{k=0}^{i-1} f'(k) \leqslant x_{n,i}$.

The other bound easily follows from the following lemma.

Lemma 1. There exists a constant b such that

$$\prod_{k=0}^{i-2} \theta_k \geqslant \frac{\Psi_{i-1, i-1}}{\Psi_{1,1}} \cdot \frac{i+1}{b^i} .$$

Proof. Given a set $S_i = \{ a_1, a_2, \ldots . a_i \}$ the number of feasible solutions builded on S_i is given by $\Psi_{i,i}$. In the following, for sake of simplicity, we will write Ψ_i instead of $\Psi_{i,i}$.

Let us define
$T_1 =$ the set of solutions $Y \in SOL(I)$ of size i such that by eliminating a_1 we will have a solution of size i-1.

Clearly $\Psi_i = | \bigcup_{1=1}^{i} T_1 |$, and $T_1 = T_m$, for every pair of indexes l, m.

We can write

$$\left| \bigcup_{l=1}^{i} T_l \right| = \sum_{l=1}^{i} |T_l| - \sum_{l_1=1}^{i-1} \sum_{l_2=l_1}^{i} |T_{l_1} \cap T_{l_2}| +$$

$$+ \sum_{l_1=1}^{i-2} \sum_{l_2=l_1}^{i-1} \sum_{l_3=l_2}^{i} |T_{l_1} \cap T_{l_2} \cap T_{l_3}|$$

$$= \left(\sum_{l=1}^{i} |T_l| - \sum_{l=2}^{i} |T_1 \cap T_l| \right) (1 - c_i)$$

for some c_i, $0 < c_i < 1$. We have by definition

$$|T_1| = \theta_{i-1} \Psi_{i-1}$$

$$\Psi_i = (i |T_1| - (i-1) |T_1 \cap T_l|) (1 - c_i)$$

$$= (i \theta_{i-1} \Psi_{i-1} - (i-1) \Psi_{i-2} \theta_{i-2} \theta_{i-1} + (i-1) \Psi_{i-2} \theta_{i-2} \theta_1)(1 - c_i)$$

$$= \theta_{i-1} (i \Psi_{i-1} - (i-1) \Psi_{i-2} \theta_{i-2} (1 - \frac{\theta_1}{\theta_{i-1}}) (1 - c_i)$$

dividing by Ψ_i we obtain

$$1 = \theta_{i-1} (i \frac{\Psi_{i-1}}{\Psi_i} - (i-1) \frac{\Psi_{i-2}}{\Psi_i} \theta_{i-2} (1 - \frac{\theta_1}{\theta_{i-1}})) (1 - c_1)$$

$$\theta_{i-2} = \frac{\Psi_{i-1}}{\Psi_{i-2}} \frac{i}{i-1} \cdot \frac{1}{c}$$

where c is a suitable quantity bounded by a constant b. Hence we obtain

$$\prod_{k=0}^{i-2} \theta_k \geqslant \frac{\Psi_{i-1}}{\Psi_1} \frac{i+1}{b^i}$$

This completes the proof of Lemma 1 and of the theorem.

Q.E.D.

Therefore Theorem 1 assures that $E_n(i)$ verifies the following inequalities

$$\binom{n}{i} \prod_{k=0}^{i-1} f(k) \leqslant E_n(i) \leqslant \binom{n}{i} e^i \prod_{k=0}^{i-1} f(k)$$

Let z_n be one of the values that maximizes $E_n(i)$, that is

$$E_n(z_n) = \max_{1 \leqslant i \leqslant n} E_n(i)$$

In the following we will assume $z_n < n^{1-\epsilon}$

The following theorem will provide bounds on $f(z_n)$ and $f(z_n\text{-}1)$

Theorem 2. Let P be a weakly hereditary maximization problem; we have

a) $\quad f(z_n-1) \geqslant \dfrac{z_n}{e(n-z_n+1)}$

b) $\quad f(z_n) \quad \leqslant \dfrac{z_n+1}{n-z_n}$

Proof a. By definition of z_n we have

$$E_n(z_n) = x_{n,z_n} \prod_{k=0}^{z_n-1} f(k) \geqslant E_n(z_n-1) = x_{n,z_n-1} \cdot \prod_{k=0}^{z_n-2} f(k)$$

$$f(z_n-1) \geqslant \frac{x_{n,z_n-1}}{x_{n,z_n}} \geqslant \frac{z_n}{e(n-z_n+1)}$$

b. In this case we have

$$E_n(z_n) = \chi_{n,z_n} \prod_{k=0}^{z_n-1} f(k) \geqslant E_n(z_n+1) = \chi_{n,z_n+1} \prod_{k=0}^{z_n} f(k)$$

$$f(z_n) \leqslant \frac{\chi_{n,z_n}}{\chi_{n,z_n+1}} \leqslant \frac{z_n+1}{n-z_n+1}$$

Q.E.D.

5. Analysis of the greedy algorithm

The heuristic we consider is the well known greedy algorithm.

Greedy algorithm

> *Input:*   an instance of WHOP with set of objects I;
>
> *begin*
>
> SOL: = ∅ ;
>
> *While* I ≠ ∅ *do*
>
>> *begin*
>>
>> choose a_j ∈ I randomly;
>>
>> *if* SOL ∪ $\{a_j\}$ is a solution
>>
>>> *then* SOL: = SOL ∪ $\{a_j\}$;
>>
>> I : = I − $\{a_j\}$
>>
>> *end*
>
> *end*
>
> *output* : $z_n^{GR} = |$ SOL $|$.

The greedy algorithm has been widely analyzed from a probabilistic point of view, because the analysis of more sophisticated heuristics becomes very difficult for stochastic dependence problems. It has been shown that for some problems this algorithm gives a good approximate solution for most cases.

Theorem 3. Let P be a weakly hereditary optimization problem. For every $\epsilon > 0$ $z_n^{GR} \geq (1-\epsilon)z_n$ almost surely.

Proof. The probability that the greedy algorithm achieves a solution of size k is greater than the probability that, for all i, $1 \leq i \leq k$, the i-th element is added to the solution in less than $\dfrac{n}{k}$ trials.

Formally we have

$$\text{Prob } (z_n^{GR} \geq (1-\epsilon)\, z_n) \geq \prod_{i=0}^{z_n(1-\epsilon)-1} (1-(1-f(i))^{\frac{n}{z_n(1-\epsilon)-1}})$$

$$\geq \prod_{i=0}^{z_n(1-\epsilon)-1} \left(1-\left(1-(f(z_n-1)\left(\frac{i}{z_n-1}\right)^{\alpha}\right)^{\frac{n}{z_n(1-\epsilon)-1}}\right)$$

where $\alpha = 1-\epsilon'$, $\epsilon' > 0$

$$\geq \prod_{i=0}^{z_n(1-\epsilon)-1} \left(1-\left(1-\frac{z_n}{(n-z_n+1)e}^{\left(\frac{i}{z_n-1}\right)^{\alpha}}\right)^{\frac{n}{z_n(1-\epsilon)-1}}\right) \quad \text{by Theorem 2}$$

$$\geq \prod_{i=0}^{z_n(1-\epsilon)-1} \left(1-e^{-\left(\frac{z_n}{(n-z_n+1)\cdot e}\right)^{\left(\frac{i}{z_n-1}\right)^{\alpha}}\cdot\frac{n}{z_n(1-\epsilon)-1}}\right)$$

since $1-x \leq e^{-x}$

$$\geq \left(1 - e^{-\left(\frac{z_n}{(n-z_n+1)e}\right)^{(1-\epsilon)^{\alpha}} \frac{n}{z_n(1-\epsilon)-1}}\right)^{z_n(1-\epsilon)} \quad \text{since} \frac{z_n}{(n-z_n+1)e} < 1$$

$$\geq \left(1 - e^{-\frac{n^{\delta n}}{(z_n(1-\epsilon)-1)^{\delta n}}}\right)^{z_n(1-\epsilon)} \qquad \delta_n < 1$$

$$\geq 1 - z_n(1-\epsilon)\, e^{-\frac{n^{\delta n}}{(z_n(1-\epsilon)-1)^{\delta n}}} \qquad \begin{array}{l} \text{since } (1-x)^m \geq 1 - mx \\ \text{for } (1-x) \geq 0 \end{array}$$

$$\geq 1 - o(n^2)$$

Q.E.D.

The theorem states a nice relationship between the solution given by the greedy algorithm and the density of the set of possible solutions. Intuitively, if we define a maximal solution as a solution that cannot be increased by adding a new element, the theorem claims that it is very unlikely that there exists a maximal solution of size less than z_n.

6. Analysis of the optimal solution

In this paragraph we will complete our analysis by evaluating the ratio z_n^* / z_n^{GR} that expresses the performance of the greedy algorithm; we will use again the combinatorial properties obtained in paragraphs 4 and 5 to derive an upper bound on z_n^*/z_n. This result, jointly with theorem 3, gives the desired bounds.

Theorem 4. Let P be a WHOP. There exists a constant d such that for n sufficiently large $\dfrac{z_n^*}{z_n^{GR}} \leqslant (d+1)^{1/d}$ almost surely .

Proof. We have that for every i

$$\text{Prob } (z_n^* \geqslant i) \leqslant E_n(i) \leqslant \binom{n}{i} e^i \prod_{k=0}^{i-1} f(k)$$

From the properties of f we obtain

$$\text{Prob } (z_n^* \geqslant i) \leqslant \binom{n}{i} e^i \prod_{k=0}^{i-1} f(z_n)^{(k/z_n)^d}$$

$$\leqslant \left(\frac{n\,e^2}{i}\right)^i \prod_{k=0}^{i-1} f(z_n)^{(k/z_n)^d} \qquad \text{since } \binom{n}{i} \leqslant \left(\frac{n\,e}{i}\right)^i$$

$$\leqslant \left(\frac{n\,e^2}{i}\right)^i \prod_{k=0}^{i-1} \left(\frac{z_n+1}{n-z_n}\right)^{(k/z_n)^d} \qquad \text{by Theorem 2}$$

$$\leqslant \left(\frac{n\,e^2}{i}\right)^i \left(\frac{z_n+1}{n-z_n}\right)^{\frac{\sum_{k=0}^{i-1} k^d}{z_n^d}}$$

$$\leqslant \left(\frac{n\,e^2}{i}\right)^i \left(\frac{z_n+1}{n-z_n}\right)^{\sum_{k=0}^{i-1} \frac{k^d}{z_n^d}}$$

Now we evaluate $\displaystyle\sum_{k=0}^{i-1} k^d$

Lemma 2. $\displaystyle\sum_{k=0}^{i-1} k^d = \frac{(i-1)^{d+1}}{d+1} + o\left((i-1)^{d+1}\right)$

Proof. Applying the Eulero Mc Laurin formula

$$\sum_{k=0}^{i-1} k^d = \int_0^{i-1} k^d \, di + \frac{1}{2}(i-1)^d - \sum_{k=1}^{d/2} [(i-1)^d]^{[2k-1]} \cdot$$

$$\cdot \frac{(-1)^k b_k}{(2k)!}$$

where $[k]$ is the k-th derivative and b_i are the Bernoulli coefficients. The thesis follows by doing some easy algebraic steps.

Now we come back to the proof of the theorem. Using the lemma and substituing $i = (1+\epsilon)(d+1)z_n$ we have for n sufficiently large

$$\text{Prob}\left\{ z_n^* \geq i \right\} \leq \left(\frac{ne^2}{i}\right)^i \left(\frac{z_n+1}{n-z_n}\right)^{\frac{i^{d+1}}{(d+1)z_n^d}}$$

$$\leq \left(\frac{ne^2}{i}\right)^i \left(\frac{2z_n}{n}\right)^{\frac{i^{d+1}}{(d+1)z_n^d}}$$

$$\leq \left(\frac{e^2}{(1+\epsilon)(d+1)^{1/d}}\right)^{(1+\epsilon)(d+1)^{1/d} \cdot z_n}$$

$$\cdot \left(\frac{z_n}{n}\right)^{-(1+\epsilon)(d+1)^{1/d} z_n + \frac{i^{(d+1)}}{(d+1)z_n^d}} \frac{i^{d+1}}{(d+1)z_n^d} \cdot 2$$

$$\leqslant \left(\frac{e^2}{(1+\epsilon)(d+1)^{1/d}}\right)^{(1+\epsilon)(d+1)^{1/d} \cdot z_n} \cdot 2^{\frac{i^{d+1}}{(d+1)\, z_n^d}} \cdot$$

$$\cdot \left(\frac{z_n}{n}\right)^{-(1+\epsilon)(d+1)^{1/d} \cdot z_n + \frac{(1+\epsilon)^{(d+1)} (d+1)^{\frac{d+1}{d}} \cdot z_n^{d+1}}{z_n^d\,(d+1)}}$$

$$\leqslant \left(\frac{e^2}{(1+\epsilon)(d+1)^{1/d}}\right)^{(1+\epsilon)(d+1)^{1/d} \cdot z_n} \cdot 2^{\frac{i^{d+1}}{(d+1)\,z_n^d}} \cdot$$

$$\cdot \left(\frac{z_n}{n}\right)^{-(1+\epsilon)(d+1)^{1/d} \cdot z_n + (1+\epsilon)^{d+1}(d+1)^{1/d} \cdot z_n}$$

$$\leqslant \left(\frac{e^2}{(1+\epsilon)(d+1)^{1/d}}\right)^{(1+\epsilon)(d+1)^{1/d} \cdot z_n} \cdot 2^{\frac{i^{d+1}}{(d+1)\,z_n^d}} \cdot$$

$$\cdot \left(\frac{z_n}{n}\right)^{(d+1)^{1/d} \cdot z_n \,\left((1+\epsilon)^{d+1} - (1+\epsilon)\right)}$$

$$\leqslant \left(\frac{e^2}{(1+\epsilon)(d+1)^{1/d}}\right)^{(1+\epsilon)(d+1)^{1/d} \cdot z_n} \cdot 2^{\frac{i^{d+1}}{(d+1)\cdot z_n^d}} \cdot$$

$$\cdot \left(\frac{z_n}{n}\right)^{(d+1)^{1/d} \cdot \epsilon \cdot z_n}$$

$$\leqslant (a) \; {}^{(1+\epsilon)^{d+1}} \; (d+1)^{1/d} \cdot z_n \; \cdot \; \left(\frac{z_n}{n}\right)^{(d+1)^{1/d}} \cdot \epsilon \cdot z_n$$

with a constant $< e^2$

$$\leqslant o(n^2) \qquad \text{since } z_n < n^{1-\epsilon} \text{ for n sufficiently large}$$

Q.E.D.

Coming back to the examples of par. 3, the problem listed in par. 2 verify the hyphoteses of the theorem

1) MAX-CLIQUE. The hypotheses of Theorem 4 is satisfied assuming $d = 1$.

    Therefore $\dfrac{z_n^*}{z_n^{GR}} \leqslant 2$ a.s. So we are able to find again the same

    result shown in [3].

2) MAX-TREE $d = 1$.

    So $\dfrac{z_n^*}{z_n^{GR}} \leqslant 2$ a.s., which is the result proved in [6].

3) K-MAX-SETPACKING $d = 1$.

    $\dfrac{z_n^*}{z_n^{GR}} \leqslant 2$ a.s. It was shown in [8].

4) K-MAX-HYPERCLIQUE. It can be easily seen that $d = K-1$.

$$\frac{z_n^*}{z_n^{GR}} \leqslant K^{\frac{1}{K-1}} \quad \text{a.s.}$$

5) MAX INDUCED PATH. Until now this problem had not been studied from a probabilistic point of view. The analysis becomes immediate applying the theorem. Also in this case $d=1$.

$$\frac{z_n^*}{z_n^{GR}} \leqslant 2 \text{ a.s.}$$

References

[1] D. Angluin, L. G. Valiant *"Fast probabilistic algorithms for hamiltonian circuits and matchings"*, Proc. IX Symposium on Theory of Computing, Boulder, (1977).

[2] M. R. Garey, D. S. Johnson *"Computers and intractability: a guide to the theory of NP–completeness"*, Freeman, San Francisco, (1978).

[3] G. R. Grimmett, C. J. H. Diarmid *"On colouring random graphs"* Math. Proc. Camb. Phil. Soc. Vol. 77 (1975).

[4] R. M. Karp *"The probabilistic analysis of some combinatorial search algorithms"* In *"Algorithms and complexity"*, J. F. Traub (ed.), Academic Press, (1976).

[5] B. Korte, L. Lovasz *"Greedoids – A structural framework for the greedy algorithm"*, Rep. n. 82230 – OR. University of Bonn, (1982).

[6] A. Marchetti–Spaccamela, M. Protasi *"The largest tree in a random graph"* Theoretical Computer Science Vol. 23 (1983).

[7] J. H. Reif, P. J. Spirakis *"Random matroids"*, Proc. XII Symposium on Theory of Computing, Los Angeles, (1980).

[8] R. Terada *"Fast algorithms for NP – hard problems which are optimal or nearoptimal with probability one"* Ph. D. Thesis, Dep. of Comp. Sci., University of Winsconsin, (1979).

[9] I. Tomescu *"Evaluation asymptotiques du nombre des cliques des hypergraphes uniformes"*, Calcolo, Vol. 18, (1981).

[10] M. Yannakakis *"The effect of a connectivity requirement on the complexity of maximum subgraph problems"*, J. ACM Vol. 26, (1979).

Annals of Discrete Mathematics 25 (1985) 311-320

A THRESHOLD FOR MULTIPLE EDGE COVERINGS IN RANDOM HYPERGRAPHS

C. VERCELLIS

Dipartimento di Matematica, Università di Milano, Milano, Italy

A threshold on the number of edges of a random hypergraph is derived for the existence of a multiple edge covering. To this end, the output of a simple probabilistic procedure is analysed in terms of convergence in law, exploiting a connection with the theory of extreme order statistics.

1. Introduction

A hypergraph $H = (V, E)$ consists of a set V of n vertices $\{v_1, v_2, \dots v_n\}$ and a collection $E = \{e_1, e_2, \dots e_m\}$ of m subsets of V called hyperedges (or simply edges). A hypergraph H is called h-uniform if all its edges have the same cardinality h: $|e_i| = h$, $i = 1,2,\dots m$.

A (multiple) edge r-covering in H is a subset $C \subseteq E$ such that each vertex of V belongs to at least r edges of the set C. The property that a hypergraph H with n vertices and m edges contains an edge r-covering will be indicated as $PC_m^r(n)$.

The two following models of random hypergraphs are considered in this paper, where $h = h(n)$ is any function of n such that $h = o(n)$ as $n \to \infty$:

(i)   the m edges of H are randomly drawn (without replacement) from the set $\Omega_h(n)$ of all the $\binom{n}{h}$ subsets of h vertices of V;

(ii) for each edge e_i, the probability that it contains any given vertex v_j is $p(n) = h/n$, independently of any other vertex or edge.

Model (i) describes h-uniform hypergraphs, and generalizes the model of random graphs considered by Erdös and Rényi [2], which corresponds to the case h=2 in (i). Model (ii) represents a more general class of hypergraphs, of which the h-uniform ones are, in a certain sense, "average" members.

For both models, a threshold function $\bar{m}^r(n)$ for the property $PC_m^r(n)$ is derived in Section 3, such that

$$Pr\left\{PC_m^r(n)\right\} \to 0 \text{ if } \frac{m(n)}{\bar{m}^r(n)} \to 0, \qquad \text{as } n \to \infty,$$

$$Pr\left\{PC_m^r(n)\right\} \to 1 \text{ if } \frac{m(n)}{\bar{m}^r(n)} \to \infty, \qquad \text{as } n \to \infty.$$

In order to obtain the expression of $\bar{m}^r(n)$, the following related problem is considered: Suppose that the edges of the hypergraph H on n vertices are chosen sequentially according to either model (i) or (ii), and let this process continue until H contains an edge r-covering. Let also $M^r(n)$ be the random variable (r.v.) representing the number of edges introduced in H before the process is stopped. Thus, the question arises of how to describe the asymptotic behaviour of the sequence $\{M^r(n)\}$ in terms of stochastic convergence as $n \to \infty$.

It is shown in Section 2 that the r.v. 's of the sequence $\{M^r(n)\}$ possess a nondegenerate limiting distribution. Specifically, it is proved that there exist normalizing sequences $\{a^r(n)\}$ and $\{b^r(n)\}$ of real numbers such that, for every real number x,

$$\lim_{n \to \infty} Pr\left\{ \frac{M^r(n) - a^r(n)}{b^r(n)} \leq x \right\} = \exp\left\{-e^{-x}\right\}. \qquad (1)$$

A connection with the theory of extreme order statistics is pointed out, enabling to derive the appropriate expressions of $a^r(n)$ and $b^r(n)$.

An alternative way of expressing the property $PC_m^1(n)$ is the following: The hypergraph H does not contain isolated vertices. With respect to this latter formulation, a threshold for $PC_m^1(n)$ has been obtained in [2] for random graphs, i.e. for the particular case h=2 of model (i).

It can be noticed that, in the case h=2, the threshold $\bar{m}^r(n)$ obtained in Section 3 reduces for r=1 to the threshold given in [2].

We finally observe that a threshold for the property PC_m^1 (n) has been derived in [4] in the case of model (ii) and h=λn, $0 < \lambda < 1$.

2. A limiting distribution for M^r (n)

Let us consider the following procedure, operating on the set V of n vertices:

```
procedure r-cover;
    begin E: = ∅;  m: = 0;
        while (PCᵐ (n) is false) do begin
            m : = m +1;
            e : = new_edge (V, E);
            E : = E ∪ {e} ;
        end;
        return (m);
    end.
```

Different versions of the procedure r-cover can be obtained by specifying different rules of choice of the edge e selected by the function new_edge for entering the set E at the m-th iteration. In particular, three alternatives are considered here:

(R1) The edge e is randomly chosen among the $\left[\binom{n}{h} - m \right]$ subsets of Ω_h (n) which have not yet been included in E;

(R2) The edge e is randomly generated in such a way that each vertex has probability p(n) of belonging to e;

(R3) The edge e is randomly chosen in the whole set Ω_h (n).

It is apparent that the hypergraphs generated through rules (R1) and (R2) are realizations of random hypergraphs according to, respectively, models (i) and (ii). The rule (R3), instead, leads to h-uniform hypergraphs which are not necessarily simple (i.e. the same edge might appear more than once).

The r.v. returned by the procedure r-cover will be denoted in general as M^r (n); the specific notation M_ℓ^r (n), $\ell=1, 2, 3$, will be used to explicitly evidentiate that rule (Rℓ) has been incorporated in the function new_edge. The asymptotic behaviour of the sequences $\{M_1^r$ (n)$\}$ and $\{M_3^r$ (n)$\}$ is the same, as shown by the following lemma, so that the analysis of convergence in law will be carried out for $\{M_3^r$ (n)$\}$ only.

Lemma 1. For the sequences of r.v. 's $\{M_1^r(n)\}$ and $\{M_3^r(n)\}$, and for any integer k,

$$\lim_{n\to\infty} \frac{\Pr\{M_1^r(n) = k\}}{\Pr\{M_3^r(n) = k\}} = 1.$$

Proof: Let Z_k (n) be the event $\{$the hypergraph consisting of the first k edges chosen by the procedure r-cover according to the rule (R3) is simple$\}$, and let $\bar{Z}_k$ (n) be its complement. Then,

$$\Pr\{M_3^r(n) = k\} = \Pr\{M_3^r(n) = k|Z_k(n)\}\Pr\{Z_k(n)\}$$
$$+ \Pr\{M_3^r(n) = k|\bar{Z}_k(n)\}\Pr\{\bar{Z}_k(n)\}.$$

We have also

$$\Pr\{Z_k(n)\} = \frac{\prod_{j=0}^{k-1}\left[\binom{n}{h} - j\right]}{\binom{n}{h}^k} \to 1 \qquad \text{as } n\to\infty.$$

The required result follows observing that, if n is large enough so that $\binom{n}{h} > k$, then $\Pr\{M_3^r(n) = k \mid Z_k(n)\} = \Pr\{M_1^r(n) = k\}$. ∎

Let W_j^r (n) be the number of the iteration at which the vertex v_j is r-covered for the first time, during an execution of the procedure r-cover. Thus, on the basis of the relation M^r (n) = max$\{W_i^r(n): 1\leq j\leq n\}$, one is naturally led to investigate the behaviour of the sequence $\{M^r(n)\}$ in the frame of the asymptotic theory of extreme order statistics.

Let us first restrict the attention to the case of rule (R2), for which the corresponding r.v. 's W_j^r (n), $1\leq j\leq n$, are independent and identically distributed, so that a nondegenerate limiting distribution takes necessarily one of the three forms described in [3]. In particular, if $r=1$, the r.v. 's W_j^1 (n) have a common geometric distribution of parameter $p(n)$, and one can easily derive from [3], Th. 2.1.3, the expressions of the normalizing constants

$$a^1 (n) = \frac{n}{h} \log n, \quad b^1 (n) = \frac{n}{h} \ ,$$

such that relation (1) holds.

In general, for $r \geq 1$, the r.v. 's W_j^r (n) have a common negative binomial distribution of parameters $r-1$ and $p=p(n)$. Again, it is possible to obtain the appropriate expressions of the normalizing constants a^r (n) and b^r (n) from results of extreme order statistics. In order to see that, let $F(t)$ be the distribution of W_j^r (n):

$$F(t) = \sum_{k=r}^{\lfloor t \rfloor} \binom{k-1}{r-1} p^r (1-p)^{k-r} , \quad t \geq r. \tag{2}$$

According to [3], we have

$$a^r (n) = \inf \{x: 1 - F(x) \leq \frac{1}{n} \}, \quad b^r(n) = L(a^r(n)),$$

where

$$L(t) = (1 - F(t))^{-1} \int_t^\infty (1 - F(y)) \, dy. \tag{3}$$

It is easy to verify that, if $x = x(n)$ is a sequence of integers such that $x(n) \cdot p(n) \to \infty$ as $n \to \infty$, then

$$1 - F(x) \sim \binom{x}{r-1} p^{r-1} (1-p)^{x-r+1} . \tag{4}$$

Combining (2), (3) and (4), on obtains the normalizing constants

$$a^r (n) = \frac{n}{h} [\log n + (r-1) \log\log n - \log(r-1)!] , \quad b^r (n) = \frac{n}{h} . \tag{5}$$

When rules (R1) and (R3) are considered in this procedure r-cover instead of (R2), the r.v. 's W_j (n) are no more independent, so that it seems difficult to obtain the corresponding expressions for a^r (n) and

b^r (n) again in the frame of extreme order statistics. Yet, it will be proved by direct arguments, in Theorem 1, that the normalizing constants given in (5) allow to show that (1) is true also in the case of rules (R1) and (R3).

Theorem 1. For the sequence of r.v. 's $\{M_\ell^r\ (n)\}$, $\ell=1, 2, 3$, and for any real number x,

$$\lim_{n\to\infty}\ \Pr\left\{\frac{M_\ell^r\ (n)\ -\ a^r\ (n)}{b^r\ (n)}\le x\right\} = \exp\ \{-e^{-x}\}, \tag{6}$$

where a^r (n) and b^r (n) are as in (5).

Proof: Let $k = \lfloor a^r\ (n) + b^r\ (n)\ x\rfloor$, and let U_ℓ^r (n, m) be the r.v. representing the number of vertices of V covered less than r times by the first m edges chosen in the procedure r-cover according to rule (Rℓ). It is clear that M_ℓ^r (n) = min $\{m: U_\ell^r$ (n, m) = o$\}$.

Let us first consider the sequence $\{M_2^r$ (n)$\}$; we have

$$\Pr\left\{\frac{M_2^r\ (n)\ -\ a^r\ (n)}{b^r\ (n)}\le x\right\}= \Pr\{M_2^r\ (n)\le k\}$$

$$= \Pr\{U_2^r\ (n, k) = 0\} = [F(k)]^n,$$

where F($\cdot$) is defined as in (2). Thus, recalling (4) and observing that

$$\lim_{n\to\infty}\ n\log\left[1-\binom{k}{r-1}\ p^{r-1}\ (1-p)^{k-r+1}\right] = -e^{-x},$$

the required result follows for $\{M_2^r$ (n)$\}$.

Consider now the sequence $\{M_3^r$ (n)$\}$. Let C_j^r (n, k) be the event "the vertex v_j is covered by at least r of the first k edges", let $\bar{C}_j^r$ (n,k) be its complement, and set B_j^r (n,k) = $\bigcap_{i=1}^{j}\ \bar{C}_i^r$ (n, k), with B_o (n,k) =1. We have that

$$\Pr\left\{\frac{M_3^r(n) - a^r(n)}{b^r(n)} \leq x\right\} = \Pr\left\{U_3^r(n, k) = 0\right\}$$

$$= \Pr\left\{\bigcap_{j=1}^{n} C_j^r(n, k)\right\} = \sum_{j=0}^{n} \binom{n}{j}(-)^j \Pr\left\{B_j^r(n,k)\right\}. \qquad (7)$$

In order to evaluate $\Pr\{B_j^r(n, k)\}$, $j=1, 2, \ldots n$, let $D_j(n, k)$ be the r.v. representing the number of edges, among the first k, which cover vertex v_j, and let $f_k(\underline{d}) = f_k(d_1, d_2, \ldots d_n)$ be their joint density. It follows that

$$\Pr\{B_j^r(n, k)\} = \sum_{\underline{d} \in D} f_k(d_1, d_2, \ldots d_n),$$

where the last summation is taken over the set $D = \{\underline{d} : d_i \leq r-1, i = 1, 2, \ldots j\}$. Suppose that hk subsets of V, of size one are randomly drawn (with repetition) from $\Omega_1(n)$; let $D_j'(n, hk)$ be the number of such 1-subsets covering vertex v_j, and $f_{hk}'(d_1, d_2, \ldots d_n)$ the corresponding joint density. It is easy to see that $f_k(\underline{d})$ and $f_{hk}'(\underline{d})$ are asymptotically equal as $n \to \infty$. In fact, it is enough to show that $\Pr\{\max_{1 \leq j \leq n} D_j'(n, hk) > k\} \to 0$ as $n \to \infty$. Indeed,

$$\Pr\{\max_{1 \leq j \leq n} D_j'(n, hk) > k\} \leq n \Pr\{D_1'(n, hk) > k\}$$

$$= n \sum_{i > k} \binom{hk}{i} \frac{1}{n^i} \left(1 - \frac{1}{n}\right)^{hk-i}$$

$$\leq n.\exp\left\{-\frac{1}{3}\left(\frac{n}{h} - 1\right)^2 \log n\right\} \to 0 \quad \text{as } n \to \infty$$

(the last inequality is obtained applying Chernoff's bound [1] on the tails of the binomial distribution). We have that

$$\sum_{\underline{d} \in D} f_{hk}'(\underline{d}) = \sum_{\underline{d} \in D} \frac{(hk)!}{d_1! \ldots d_n!} \cdot \frac{1}{n^{hk}}$$

$$= \sum_{d_1 + \ldots d_j = t} \frac{(hk)!}{d_1! \ldots d_j!(hk-t)!} \cdot \frac{1}{n^t} \left(1 - \frac{j}{n}\right)^{hk-t},$$

with $0 \leq t \leq j(r-1)$. Moreover,

$$\binom{n}{j} \frac{(hk)\,!}{d_1!\dots d_j!\,(hk\text{-}t)\,!} \cdot \frac{1}{n^t} \cdot \left(1\text{-}\frac{j}{n}\right)^{hk\text{-}t}$$

$$\sim \frac{e^{-jx}\,[(r\text{-}1)!]^{\,j}\,(\log n)^{t-j(r-1)}}{j!\,d_1!\dots d_j!} \,. \tag{8}$$

It is easy to see that the last term in (8) tends to zero for $t < j(r\text{-}1)$, while it tends to $\dfrac{e^{-jx}}{j!}$ for $t = j(r\text{-}1)$. We obtain therefore

$$\lim_{n\to\infty} \binom{n}{j} \Pr\{B_j^r(n,k)\} = -\frac{e^{-jx}}{j!}\,,$$

which, combined with (7), leads to

$$\lim_{n\to\infty} \Pr\left\{\frac{M_3^r(n) - a^r(n)}{b^r(n)} \le x\right\} = \sum_{j=0}^{\infty} (-)^j \frac{e^{-jx}}{j!} = \exp\{-e^{-x}\}.$$

Finally, notice that (6) is true also for $\ell=2$, because of Lemma 1 and the previous result for the-case $\ell=3$. ∎

3. A threshold for multiple edge coverings

Let $\omega(n)$ be any function of n such that $\omega(n) \to \infty$ as $n\to\infty$. We have the following

Theorem 2: For both models (i) and (ii):

(a)   if $m = m(n) = (n/h)\,[\log n + (r\text{-}1) \log \log n - \log(r\text{-}1)\,! - \omega(n)]$, then

$$\lim_{n\to\infty} \Pr\{PC_m^r(n)\} = 0;$$

b)   if $m = m(n) = (n/h)\,[\log n + (r\text{-}1) \log \log n - \log(r\text{-}1)! + \omega(n)]$, then

$$\lim_{n \to \infty} \Pr\left\{PC_m^r (n)\right\} = 1$$

Proof: The m edges of the random hypergraph can be thought as being sequentially generated according to rules (R1) and (R2), respectively, for models (i) and (ii). Thus, the procedure r-cover can be applied to the resulting hypergraph, and Theorem 1 leads to the required result. ■

As both expressions for m in (a) and (b) are asymptotic to (n/h) [log n + (r-1) log log n − log (r-1)!], we finally obtain the

Corollary: For both models (i) and (ii), the threshold function for the property PC_m^r (n) is

$$\bar{m}^r(n) = (n/h) \left[\log n + (r\text{-}1) \log \log n - \log (r\text{-}1)!\right].$$

References

[1]   D. Angluin and L.G. Valiant, Fast probabilistic algorithms for Hamiltonian circuits and matchings, *J. Comput. System Sci.* 18 (1979), 155-193.

[2]   P. Erdös and A. Rényi, On random graphs I, *Publ. Math. Debrecen* 6 (1959), 290-297.

[3]   J. Galambos, *The asymptotic theory of extreme order statistics*, Wiley, New York, 1978.

[4]   C. Vercellis, A probabilistic analysis of the set covering problem, *Annals of Operations Research* 1 (to appear).

ANNALS OF DISCRETE MATHEMATICS